Essentials of Genetics

Essentials of Genetics

Edited by
Ayden Llyod

Larsen & Keller
www.larsen-keller.com

Essentials of Genetics
Edited by Ayden Llyod
ISBN: 978-1-63549-132-6 (Hardback)

Larsen & Keller
Published by Larsen and Keller Education,
5 Penn Plaza,
19th Floor,
New York, NY 10001, USA

Cataloging-in-Publication Data

Essentials of genetics / edited by Ayden Llyod.
         p. cm.
Includes bibliographical references and index.
ISBN 978-1-63549-132-6
1. Genetics. 2. Human genetics.
I. Llyod, Ayden.
QH430 .E87 2017
576.5--dc23

The publisher's policy is to use permanent paper from mills that operate a sustainable forestry policy. Furthermore, the publisher ensures that the text paper and cover boards used have met acceptable environmental accreditation standards.

Printed and bound in the United States of America.

For more information regarding Larsen and Keller Education and its products, please visit the publisher's website www.larsen-keller.com

Table of Contents

Preface

This book outlines the processes and applications of genetics in detail. It elaborates on the various techniques and studies related to the subject. Genetics, as a sub-field of biology, refers to the study of genes, heredity and genetic variation in living beings like plants, bacteria, humans and animals, etc. The topics included in the text on genetics are of utmost significance and are bound to provide incredible insights to readers. Included in this textbook are the fundamental application of genetics and the technological advances that have made this possible. It will serve as a valuable source of reference for those interested in this field.

Given below is the chapter wise description of the book:

Chapter 1- Genetics is the study of genes, heredity and genetic variation in living organisms. This chapter introduces the reader to the science of genetics and provides a brief overview of the topic. The chapter offers an insightful focus, keeping in mind the complex subject matter.

Chapter 2- To understand the complex science of genetics, it is of utmost importance that one grasps its key building blocks. The major components of genetics like gene, heredity, DNA, RNA, chromosome, DNA sequencing, and nucleotide have been discussed in the chapter. This enables the reader to form a comprehensive perception of genetics.

Chapter 3- The concepts of genetic variation, genetic drift, deletion, mutation and natural selection explain the physical and cognitive differences that exist in the human population at any given time. They also explain the reason that certain characteristics survive and others go extinct. This chapter discusses these key concepts in depth providing the reader with an awareness of the reasons behind why genetic research is one of the most rapidly growing fields of pure science.

Chapter 4- With the breakthroughs in genetics that have been a direct result of technological advancement, it comes as no surprise that the euchromatic areas of the human genome have been completely mapped by The Human Genome Project (HGP). This has enabled scientists around the globe to better understand genetic disorders and to also pursue improved gene therapy. This chapter focuses on the leaps and bounds achieved by human genetics and equips the reader with detailed information about the cutting-edge developments in human genetics.

Chapter 5- This chapter chronicles the history of genetics from the days of Mendelian or classical genetics to the current genetic research in cancer therapy. The reader is provided with in-depth information about the history of genetics from its humble beginnings at a monastery. The timeline of genetics illustrates the immense and rapid advancement the field has experienced and discusses its future.

At the end, I would like to thank all those who dedicated their time and efforts for the successful completion of this book. I also wish to convey my gratitude towards my friends and family who supported me at every step.

Editor

Introduction to Genetics

Genetics is the study of genes, heredity and genetic variation in living organisms. This chapter introduces the reader to the science of genetics and provides a brief overview of the topic. The chapter offers an insightful focus, keeping in mind the complex subject matter.

Genetics is the study of genes, genetic variation, and heredity in living organisms. It is generally considered a field of biology, but it intersects frequently with many of the life sciences and is strongly linked with the study of information systems.

The father of genetics is Gregor Mendel, a late 19th-century scientist and Augustinian friar. Mendel studied 'trait inheritance', patterns in the way traits were handed down from parents to offspring. He observed that organisms (pea plants) inherit traits by way of discrete "units of inheritance". This term, still used today, is a somewhat ambiguous definition of what is referred to as a gene.

Trait inheritance and molecular inheritance mechanisms of genes are still primary principles of genetics in the 21st century, but modern genetics has expanded beyond inheritance to studying the function and behavior of genes. Gene structure and function, variation, and distribution are studied within the context of the cell, the organism (e.g. dominance) and within the context of a population. Genetics has given rise to a number of sub-fields including epigenetics and population genetics. Organisms studied within the broad field span the domain of life, including bacteria, plants, animals, and humans.

Genetic processes work in combination with an organism's environment and experiences to influence development and behavior, often referred to as nature versus nurture. The intra- or extra-cellular environment of a cell or organism may switch gene transcription on or off. A classic example is two seeds of genetically identical corn, one placed in a temperate climate and one in an arid climate. While the average height of the two corn stalks may be genetically determined to be equal, the one in the arid climate only grows to half the height of the one in the temperate climate due to lack of water and nutrients in its environment.

Etymology

The word genetics stems from the Ancient Greek *genetikos* meaning "genitive"/"generative", which in turn derives from *genesis* meaning "origin".

The Gene

The modern working definition of a gene is a portion (or sequence) of DNA that codes for a known cellular function or process (e.g. the function "make melanin molecules"). A single 'gene' is most similar to a single 'word' in the English language. The nucleotides (molecules) that make up genes can be seen as 'letters' in the English language. Nucleotides are named according to which of the four nitrogenous bases they contain. The four bases are cytosine, guanine, adenine, and thymine. A single gene may have a small number of nucleotides or a large number of nucleotides, in the same way that a word may be small or large (e.g. 'cell' vs. 'electrophysiology'). A single gene often interacts with neighboring genes to produce a cellular function and can even be ineffectual without those neighboring genes. This can be seen in the same way that a 'word' may have meaning only in the context of a 'sentence.' A series of nucleotides can be put together without forming a gene (non coding regions of DNA), like a string of letters can be put together without forming a word (e.g. udkslk). Nonetheless, all words have letters, like all genes must have nucleotides.

A quick heuristic that is often used (but not always true) is "one gene, one protein" meaning a singular gene codes for a singular protein type in a cell (enzyme, transcription factor, etc.).

The sequence of nucleotides in a gene is read and translated by a cell to produce a chain of amino acids which in turn folds into a protein. The order of amino acids in a protein corresponds to the order of nucleotides in the gene. This relationship between nucleotide sequence and amino acid sequence is known as the genetic code. The amino acids in a protein determine how it folds into its unique three-dimensional shape, a structure that is ultimately responsible for the protein's function. Proteins carry out many of the functions needed for cells to live. A change to the DNA in a gene can alter a protein's amino acid sequence, thereby changing its shape and function and rendering the protein ineffective or even malignant (e.g. sickle cell anemia). Changes to genes are called mutations.

History

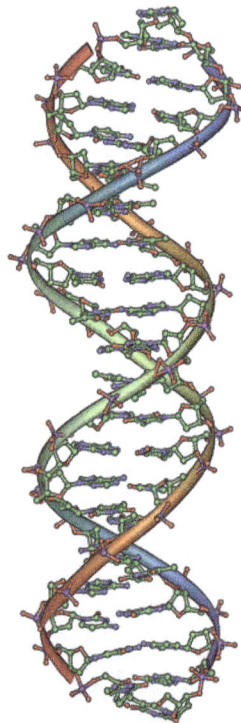

DNA, the molecular basis for biological inheritance. Each strand of DNA is a chain of nucleotides, matching each other in the center to form what look like rungs on a twisted ladder.

The observation that living things inherit traits from their parents has been used since prehistoric times to improve crop plants and animals through selective breeding. The modern science of genetics, seeking to understand this process, began with the work of Gregor Mendel in the mid-19th century.

Prior to Mendel, Imre Festetics, a Hungarian noble, who lived in Kőszeg before Gregor Mendel, was the first who used the word "genetics". He described several rules of genetic inheritance in his work *The genetic law of the Nature* (Die genetische Gesätze der Natur, 1819). His second law is the same as what Mendel published. In his third law, he developed the basic principles of mutation (he can be considered a forerunner of Hugo de Vries.)

Other theories of inheritance preceded his work. A popular theory during Mendel's time was the concept of blending inheritance: the idea that individuals inherit a smooth blend of traits from their parents. Mendel's work provided examples where traits were definitely not blended after hybridization, showing that traits are produced by combinations of distinct genes rather than a continuous blend. Blending of traits in the progeny is now explained by the action of multiple genes with quantitative effects. Another theory that had some support at that time was the inheritance of acquired characteristics: the belief that individuals inherit traits strengthened by their parents. This theory (commonly associated with Jean-Baptiste Lamarck) is now known to be wrong—the experiences of individuals do not affect the genes they pass to their children, although evidence in the field of epigenetics has revived some aspects of Lamarck's theory. Other theories included the pangenesis of Charles Darwin (which had both acquired and inherited aspects) and Francis Galton's reformulation of pangenesis as both particulate and inherited.

Mendelian and Classical Genetics

Modern genetics started with Gregor Johann Mendel, a scientist and Augustinian friar who studied the nature of inheritance in plants. In his paper "*Versuche über Pflanzenhybriden*" ("Experiments on Plant Hybridization"), presented in 1865 to the *Naturforschender Verein* (Society for Research in Nature) in Brünn, Mendel traced the inheritance patterns of certain traits in pea plants and described them mathematically. Although this pattern of inheritance could only be observed for a few traits, Mendel's work suggested that heredity was particulate, not acquired, and that the inheritance patterns of many traits could be explained through simple rules and ratios.

The importance of Mendel's work did not gain wide understanding until the 1890s, after his death, when other scientists working on similar problems re-discovered his research. William Bateson, a proponent of Mendel's work, coined the word *genetics* in 1905. (The adjective *genetic*, derived from the Greek word *genesis*—"origin", predates the noun and was first used in a biological sense in 1860.) Bateson both acted as a mentor and was aided significantly by the work of women scientists from Newnham College at Cambridge, specifically the work of Becky Saunders, Nora Darwin Barlow, and Muriel Wheldale Onslow. Bateson popularized the usage of the word *genetics* to describe the study of inheritance in his inaugural address to the Third International Conference on Plant Hybridization in London, England, in 1906.

After the rediscovery of Mendel's work, scientists tried to determine which molecules in the cell were responsible for inheritance. In 1911, Thomas Hunt Morgan argued that genes are on chromosomes, based on observations of a sex-linked white eye mutation in fruit flies. In 1913, his student

Alfred Sturtevant used the phenomenon of genetic linkage to show that genes are arranged linearly on the chromosome.

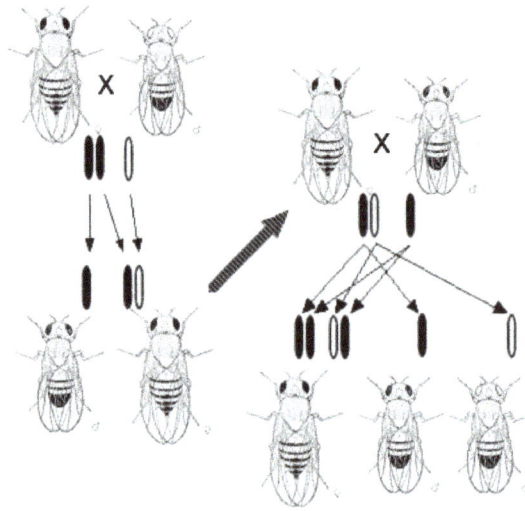

Morgan's observation of sex-linked inheritance of a mutation causing white eyes in Drosophila led him to the hypothesis that genes are located upon chromosomes.

Molecular Genetics

Although genes were known to exist on chromosomes, chromosomes are composed of both protein and DNA, and scientists did not know which of the two is responsible for inheritance. In 1928, Frederick Griffith discovered the phenomenon of transformation dead bacteria could transfer genetic material to "transform" other still-living bacteria. Sixteen years later, in 1944, the Avery–MacLeod–McCarty experiment identified DNA as the molecule responsible for transformation. The role of the nucleus as the repository of genetic information in eukaryotes had been established by Hämmerling in 1943 in his work on the single celled alga *Acetabularia*. The Hershey–Chase experiment in 1952 confirmed that DNA (rather than protein) is the genetic material of the viruses that infect bacteria, providing further evidence that DNA is the molecule responsible for inheritance.

James Watson and Francis Crick determined the structure of DNA in 1953, using the X-ray crystallography work of Rosalind Franklin and Maurice Wilkins that indicated DNA had a helical structure (i.e., shaped like a corkscrew). Their double-helix model had two strands of DNA with the nucleotides pointing inward, each matching a complementary nucleotide on the other strand to form what looks like rungs on a twisted ladder. This structure showed that genetic information exists in the sequence of nucleotides on each strand of DNA. The structure also suggested a simple method for replication: if the strands are separated, new partner strands can be reconstructed for each based on the sequence of the old strand. This property is what gives DNA its semi-conservative nature where one strand of new DNA is from an original parent strand.

Although the structure of DNA showed how inheritance works, it was still not known how DNA influences the behavior of cells. In the following years, scientists tried to understand how DNA controls the process of protein production. It was discovered that the cell uses DNA as a template to create matching messenger RNA, molecules with nucleotides very similar to DNA. The nucleotide sequence of a messenger RNA is used to create an amino acid sequence in protein; this translation between nucleotide sequences and amino acid sequences is known as the genetic code.

With the newfound molecular understanding of inheritance came an explosion of research. A notable theory arose from Tomoko Ohta in 1973 with her amendment to the neutral theory of molecular evolution through publishing the nearly neutral theory of molecular evolution. In this theory, Ohta stressed the importance of natural selection and the environment to the rate at which genetic evolution occurs. One important development was chain-termination DNA sequencing in 1977 by Frederick Sanger. This technology allows scientists to read the nucleotide sequence of a DNA molecule. In 1983, Kary Banks Mullis developed the polymerase chain reaction, providing a quick way to isolate and amplify a specific section of DNA from a mixture. The efforts of the Human Genome Project, Department of Energy, NIH, and parallel private efforts by Celera Genomics led to the sequencing of the human genome in 2003.

Features of Inheritance

Discrete inheritance and Mendel's laws

A Punnett square depicting a cross between two pea plants heterozygous for purple (B) and white (b) blossoms.

At its most fundamental level, inheritance in organisms occurs by passing discrete heritable units, called genes, from parents to progeny. This property was first observed by Gregor Mendel, who studied the segregation of heritable traits in pea plants. In his experiments studying the trait for flower color, Mendel observed that the flowers of each pea plant were either purple or white—but never an intermediate between the two colors. These different, discrete versions of the same gene are called alleles.

In the case of the pea, which is a diploid species, each individual plant has two copies of each gene, one copy inherited from each parent. Many species, including humans, have this pattern of inheritance. Diploid organisms with two copies of the same allele of a given gene are called homozygous at that gene locus, while organisms with two different alleles of a given gene are called heterozygous.

The set of alleles for a given organism is called its genotype, while the observable traits of the organism are called its phenotype. When organisms are heterozygous at a gene, often one allele is called dominant as its qualities dominate the phenotype of the organism, while the other allele is called recessive as its qualities recede and are not observed. Some alleles do not have complete dominance and instead have incomplete dominance by expressing an intermediate phenotype, or codominance by expressing both alleles at once.

When a pair of organisms reproduce sexually, their offspring randomly inherit one of the two alleles from each parent. These observations of discrete inheritance and the segregation of alleles are collectively known as Mendel's first law or the Law of Segregation.

Notation and Diagrams

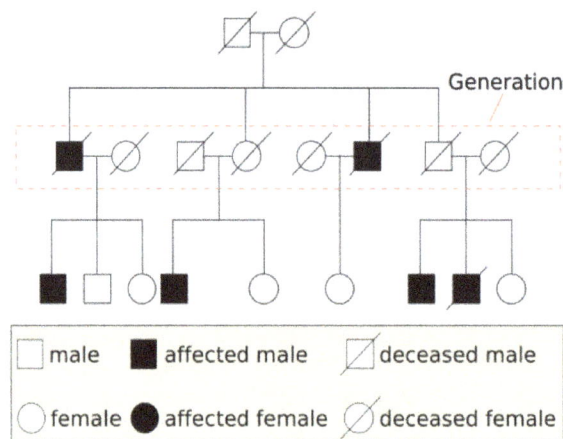

Genetic pedigree charts help track the inheritance patterns of traits.

Geneticists use diagrams and symbols to describe inheritance. A gene is represented by one or a few letters. Often a "+" symbol is used to mark the usual, non-mutant allele for a gene.

In fertilization and breeding experiments (and especially when discussing Mendel's laws) the parents are referred to as the "P" generation and the offspring as the "F1" (first filial) generation. When the F1 offspring mate with each other, the offspring are called the "F2" (second filial) generation. One of the common diagrams used to predict the result of cross-breeding is the Punnett square.

When studying human genetic diseases, geneticists often use pedigree charts to represent the inheritance of traits. These charts map the inheritance of a trait in a family tree.

Multiple Gene Interactions

Organisms have thousands of genes, and in sexually reproducing organisms these genes generally assort independently of each other. This means that the inheritance of an allele for yellow or green pea color is unrelated to the inheritance of alleles for white or purple flowers. This phenomenon, known as "Mendel's second law" or the "Law of independent assortment", means that the alleles of different genes get shuffled between parents to form offspring with many different combinations. (Some genes do not assort independently, demonstrating genetic linkage, a topic discussed later in this article.)

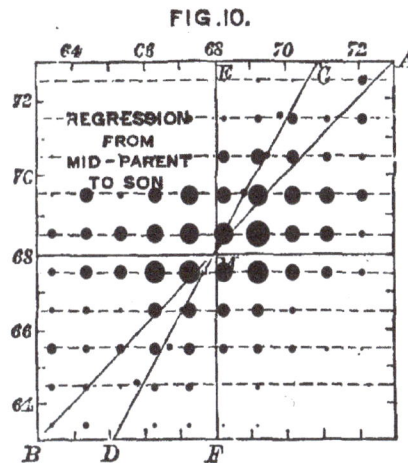

Human height is a trait with complex genetic causes. Francis Galton's data from 1889 shows the relationship between offspring height as a function of mean parent height.

Often different genes can interact in a way that influences the same trait. In the Blue-eyed Mary (*Omphalodes verna*), for example, there exists a gene with alleles that determine the color of flowers: blue or magenta. Another gene, however, controls whether the flowers have color at all or are white. When a plant has two copies of this white allele, its flowers are white—regardless of whether the first gene has blue or magenta alleles. This interaction between genes is called epistasis, with the second gene epistatic to the first.

Many traits are not discrete features (e.g. purple or white flowers) but are instead continuous features (e.g. human height and skin color). These complex traits are products of many genes. The influence of these genes is mediated, to varying degrees, by the environment an organism has experienced. The degree to which an organism's genes contribute to a complex trait is called heritability. Measurement of the heritability of a trait is relative—in a more variable environment, the environment has a bigger influence on the total variation of the trait. For example, human height is a trait with complex causes. It has a heritability of 89% in the United States. In Nigeria, however, where people experience a more variable access to good nutrition and health care, height has a heritability of only 62%.

Molecular Basis for Inheritance

DNA and Chromosomes

The molecular basis for genes is deoxyribonucleic acid (DNA). DNA is composed of a chain of nucleotides, of which there are four types: adenine (A), cytosine (C), guanine (G), and thymine (T). Genetic information exists in the sequence of these nucleotides, and genes exist as stretches of sequence along the DNA chain. Viruses are the only exception to this rule—sometimes viruses use the very similar molecule, RNA, instead of DNA as their genetic material. Viruses cannot reproduce without a host and are unaffected by many genetic processes, so tend not to be considered living organisms.

DNA normally exists as a double-stranded molecule, coiled into the shape of a double helix. Each nucleotide in DNA preferentially pairs with its partner nucleotide on the opposite strand: A pairs with T, and C pairs with G. Thus, in its two-stranded form, each strand effectively contains all necessary information, redundant with its partner strand. This structure of DNA is the physical basis for inheritance: DNA replication duplicates the genetic information by splitting the strands and using each strand as a template for synthesis of a new partner strand.

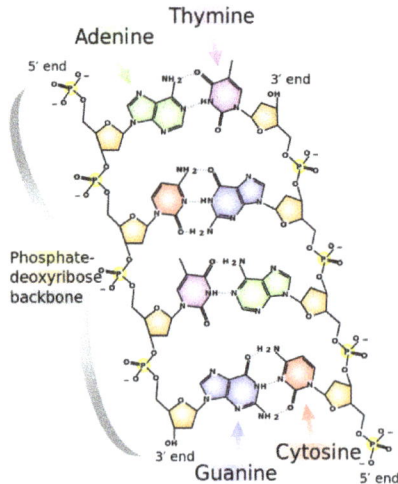

The molecular structure of DNA. Bases pair through the arrangement of hydrogen bonding between the strands.

Genes are arranged linearly along long chains of DNA base-pair sequences. In bacteria, each cell usually contains a single circular genophore, while eukaryotic organisms (such as plants and animals) have their DNA arranged in multiple linear chromosomes. These DNA strands are often extremely long; the largest human chromosome, for example, is about 247 million base pairs in length. The DNA of a chromosome is associated with structural proteins that organize, compact and control access to the DNA, forming a material called chromatin; in eukaryotes, chromatin is usually composed of nucleosomes, segments of DNA wound around cores of histone proteins. The full set of hereditary material in an organism (usually the combined DNA sequences of all chromosomes) is called the genome.

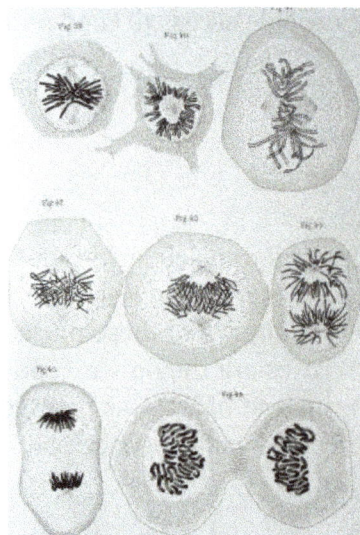

Walther Flemming's 1882 diagram of eukaryotic cell division. Chromosomes are copied, condensed, and organized. Then, as the cell divides, chromosome copies separate into the daughter cells.

While haploid organisms have only one copy of each chromosome, most animals and many plants are diploid, containing two of each chromosome and thus two copies of every gene. The two alleles for a gene are located on identical loci of the two homologous chromosomes, each allele inherited from a different parent.

Many species have so-called sex chromosomes that determine the gender of each organism. In humans and many other animals, the Y chromosome contains the gene that triggers the development of the specifically male characteristics. In evolution, this chromosome has lost most of its content and also most of its genes, while the X chromosome is similar to the other chromosomes and contains many genes. The X and Y chromosomes form a strongly heterogeneous pair.

Reproduction

When cells divide, their full genome is copied and each daughter cell inherits one copy. This process, called mitosis, is the simplest form of reproduction and is the basis for asexual reproduction. Asexual reproduction can also occur in multicellular organisms, producing offspring that inherit their genome from a single parent. Offspring that are genetically identical to their parents are called clones.

Eukaryotic organisms often use sexual reproduction to generate offspring that contain a mixture of genetic material inherited from two different parents. The process of sexual reproduction alternates between forms that contain single copies of the genome (haploid) and double copies (diploid). Haploid cells fuse and combine genetic material to create a diploid cell with paired chromosomes. Diploid organisms form haploids by dividing, without replicating their DNA, to create daughter cells that randomly inherit one of each pair of chromosomes. Most animals and many plants are diploid for most of their lifespan, with the haploid form reduced to single cell gametes such as sperm or eggs.

Although they do not use the haploid/diploid method of sexual reproduction, bacteria have many methods of acquiring new genetic information. Some bacteria can undergo conjugation, transferring a small circular piece of DNA to another bacterium. Bacteria can also take up raw DNA fragments found in the environment and integrate them into their genomes, a phenomenon known as transformation. These processes result in horizontal gene transfer, transmitting fragments of genetic information between organisms that would be otherwise unrelated.

Recombination and Genetic Linkage

The diploid nature of chromosomes allows for genes on different chromosomes to assort independently or be separated from their homologous pair during sexual reproduction wherein haploid gametes are formed. In this way new combinations of genes can occur in the offspring of a mating pair. Genes on the same chromosome would theoretically never recombine. However, they do via the cellular process of chromosomal crossover. During crossover, chromosomes exchange stretches of DNA, effectively shuffling the gene alleles between the chromosomes. This process of chromosomal crossover generally occurs during meiosis, a series of cell divisions that creates haploid cells.

The first cytological demonstration of crossing over was performed by Harriet Creighton and Bar-

bara McClintock in 1931. Their research and experiments on corn provided cytological evidence for the genetic theory that linked genes on paired chromosomes do in fact exchange places from one homolog to the other.

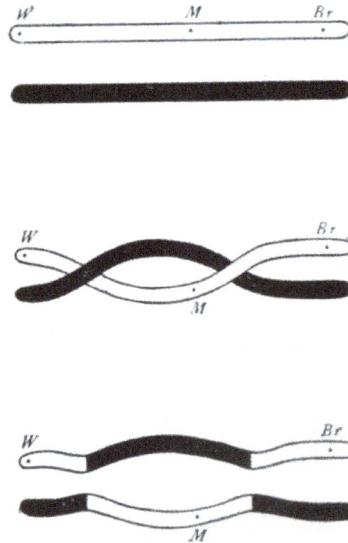

Thomas Hunt Morgan's 1916 illustration of a double crossover between chromosomes.

The probability of chromosomal crossover occurring between two given points on the chromosome is related to the distance between the points. For an arbitrarily long distance, the probability of crossover is high enough that the inheritance of the genes is effectively uncorrelated. For genes that are closer together, however, the lower probability of crossover means that the genes demonstrate genetic linkage; alleles for the two genes tend to be inherited together. The amounts of linkage between a series of genes can be combined to form a linear linkage map that roughly describes the arrangement of the genes along the chromosome.

Gene Expression

Genetic Code

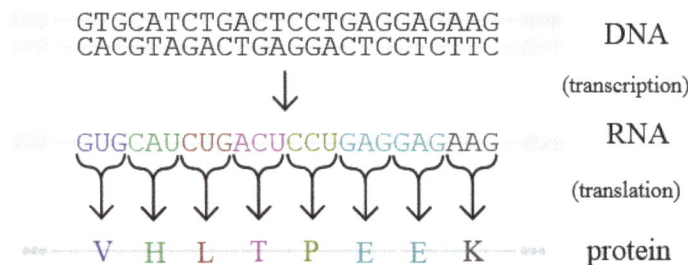

The genetic code: Using a triplet code, DNA, through a messenger RNA intermediary, specifies a protein.

Genes generally express their functional effect through the production of proteins, which are complex molecules responsible for most functions in the cell. Proteins are made up of one or more polypeptide chains, each of which is composed of a sequence of amino acids, and the DNA sequence of a gene (through an RNA intermediate) is used to produce a specific amino acid sequence. This process begins with the production of an RNA molecule with a sequence matching the gene's DNA sequence, a process called transcription.

This messenger RNA molecule is then used to produce a corresponding amino acid sequence through a process called translation. Each group of three nucleotides in the sequence, called a codon, corresponds either to one of the twenty possible amino acids in a protein or an instruction to end the amino acid sequence; this correspondence is called the genetic code. The flow of information is unidirectional: information is transferred from nucleotide sequences into the amino acid sequence of proteins, but it never transfers from protein back into the sequence of DNA—a phenomenon Francis Crick called the central dogma of molecular biology.

The specific sequence of amino acids results in a unique three-dimensional structure for that protein, and the three-dimensional structures of proteins are related to their functions. Some are simple structural molecules, like the fibers formed by the protein collagen. Proteins can bind to other proteins and simple molecules, sometimes acting as enzymes by facilitating chemical reactions within the bound molecules (without changing the structure of the protein itself). Protein structure is dynamic; the protein hemoglobin bends into slightly different forms as it facilitates the capture, transport, and release of oxygen molecules within mammalian blood.

A single nucleotide difference within DNA can cause a change in the amino acid sequence of a protein. Because protein structures are the result of their amino acid sequences, some changes can dramatically change the properties of a protein by destabilizing the structure or changing the surface of the protein in a way that changes its interaction with other proteins and molecules. For example, sickle-cell anemia is a human genetic disease that results from a single base difference within the coding region for the β-globin section of hemoglobin, causing a single amino acid change that changes hemoglobin's physical properties. Sickle-cell versions of hemoglobin stick to themselves, stacking to form fibers that distort the shape of red blood cells carrying the protein. These sickle-shaped cells no longer flow smoothly through blood vessels, having a tendency to clog or degrade, causing the medical problems associated with this disease.

Some DNA sequences are transcribed into RNA but are not translated into protein products—such RNA molecules are called non-coding RNA. In some cases, these products fold into structures which are involved in critical cell functions (e.g. ribosomal RNA and transfer RNA). RNA can also have regulatory effects through hybridization interactions with other RNA molecules (e.g. microRNA).

Nature and Nurture

Although genes contain all the information an organism uses to function, the environment plays an important role in determining the ultimate phenotypes an organism displays. This is the complementary relationship often referred to as "nature and nurture". The phenotype of an organism depends on the interaction of genes and the environment. An interesting example is the coat coloration of the Siamese cat. In this case, the body temperature of the cat plays the role of the environment. The cat's genes code for dark hair, thus the hair-producing cells in the cat make cellular proteins resulting in dark hair. But these dark hair-producing proteins are sensitive to temperature (i.e. have a mutation causing temperature-sensitivity) and denature in higher-temperature environments, failing to produce dark-hair pigment in areas where the cat has a higher body temperature. In a low-temperature environment, however, the protein's structure is stable and produces dark-hair pigment normally. The protein remains functional in areas of skin that are colder – such as its legs, ears, tail and face – so the cat has dark-hair at its extremities.

Siamese cats have a temperature-sensitive pigment-production mutation.

Environment plays a major role in effects of the human genetic disease phenylketonuria. The mutation that causes phenylketonuria disrupts the ability of the body to break down the amino acid phenylalanine, causing a toxic build-up of an intermediate molecule that, in turn, causes severe symptoms of progressive mental retardation and seizures. However, if someone with the phenylketonuria mutation follows a strict diet that avoids this amino acid, they remain normal and healthy.

A popular method in determining how genes and environment ("nature and nurture") contribute to a phenotype is by studying identical and fraternal twins or siblings of multiple births. Because identical siblings come from the same zygote, they are genetically the same. Fraternal twins are as genetically different from one another as normal siblings. By analyzing statistics on how often a twin of a set has a certain disorder compared to other sets of twins, scientists can determine whether that disorder is caused by genetic or environmental factors (i.e. whether it has 'nature' or 'nurture' causes). One famous example is the multiple birth study of the Genain quadruplets, who were identical quadruplets all diagnosed with schizophrenia.

Gene Regulation

The genome of a given organism contains thousands of genes, but not all these genes need to be active at any given moment. A gene is expressed when it is being transcribed into mRNA and there exist many cellular methods of controlling the expression of genes such that proteins are produced only when needed by the cell. Transcription factors are regulatory proteins that bind to DNA, either promoting or inhibiting the transcription of a gene. Within the genome of *Escherichia coli* bacteria, for example, there exists a series of genes necessary for the synthesis of the amino acid tryptophan. However, when tryptophan is already available to the cell, these genes for tryptophan synthesis are no longer needed. The presence of tryptophan directly affects the activity of the genes—tryptophan molecules bind to the tryptophan repressor (a transcription factor), changing the repressor's structure such that the repressor binds to the genes. The tryptophan repressor blocks the transcription and expression of the genes, thereby creating negative feedback regulation of the tryptophan synthesis process.

Differences in gene expression are especially clear within multicellular organisms, where cells all contain the same genome but have very different structures and behaviors due to the expression of different sets of genes. All the cells in a multicellular organism derive from a single cell, differentiating into variant cell types in response to external and intercellular signals and gradually establishing different patterns of gene expression to create different behaviors. As no single gene is responsible for the development of structures within multicellular organisms, these patterns arise from the complex interactions between many cells.

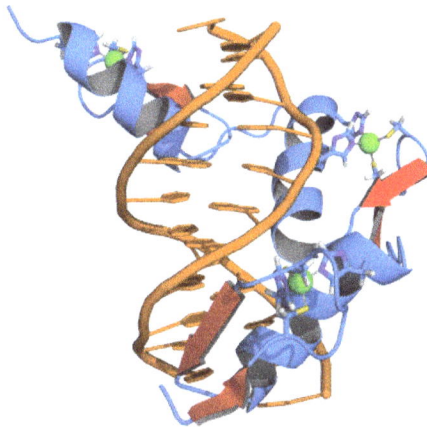

Transcription factors bind to DNA, influencing the transcription of associated genes.

Within eukaryotes, there exist structural features of chromatin that influence the transcription of genes, often in the form of modifications to DNA and chromatin that are stably inherited by daughter cells. These features are called "epigenetic" because they exist "on top" of the DNA sequence and retain inheritance from one cell generation to the next. Because of epigenetic features, different cell types grown within the same medium can retain very different properties. Although epigenetic features are generally dynamic over the course of development, some, like the phenomenon of paramutation, have multigenerational inheritance and exist as rare exceptions to the general rule of DNA as the basis for inheritance.

Genetic Change

Mutations

During the process of DNA replication, errors occasionally occur in the polymerization of the second strand. These errors, called mutations, can affect the phenotype of an organism, especially if they occur within the protein coding sequence of a gene. Error rates are usually very low—1 error in every 10–100 million bases—due to the "proofreading" ability of DNA polymerases. Processes that increase the rate of changes in DNA are called mutagenic: mutagenic chemicals promote errors in DNA replication, often by interfering with the structure of base-pairing, while UV radiation induces mutations by causing damage to the DNA structure. Chemical damage to DNA occurs naturally as well and cells use DNA repair mechanisms to repair mismatches and breaks. The repair does not, however, always restore the original sequence.

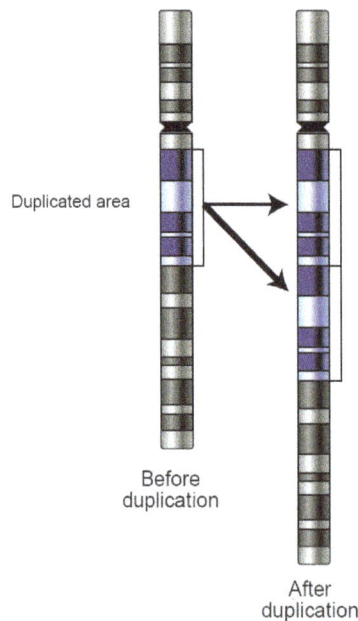

Gene duplication allows diversification by providing redundancy: one gene can mutate and lose its original function without harming the organism.

In organisms that use chromosomal crossover to exchange DNA and recombine genes, errors in alignment during meiosis can also cause mutations. Errors in crossover are especially likely when similar sequences cause partner chromosomes to adopt a mistaken alignment; this makes some regions in genomes more prone to mutating in this way. These errors create large structural changes in DNA sequence – duplications, inversions, deletions of entire regions – or the accidental exchange of whole parts of sequences between different chromosomes (chromosomal translocation).

Natural Selection and Evolution

Mutations alter an organism's genotype and occasionally this causes different phenotypes to appear. Most mutations have little effect on an organism's phenotype, health, or reproductive fitness. Mutations that do have an effect are usually deleterious, but occasionally some can be beneficial. Studies in the fly *Drosophila melanogaster* suggest that if a mutation changes a protein produced by a gene, about 70 percent of these mutations will be harmful with the remainder being either neutral or weakly beneficial.

Population genetics studies the distribution of genetic differences within populations and how these distributions change over time. Changes in the frequency of an allele in a population are mainly influenced by natural selection, where a given allele provides a selective or reproductive advantage to the organism, as well as other factors such as mutation, genetic drift, genetic draft, artificial selection and migration.

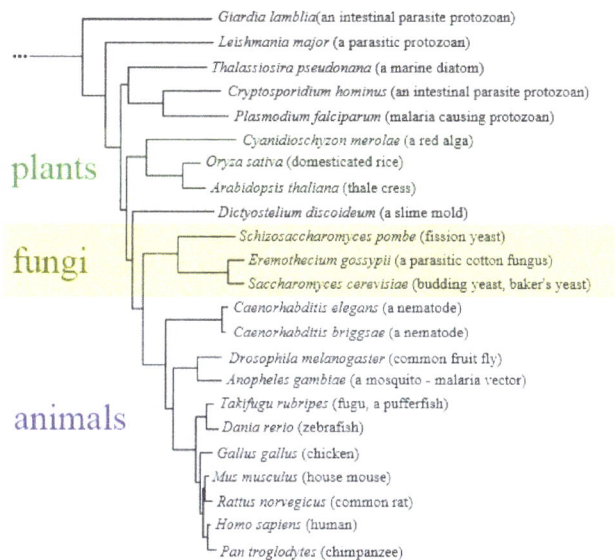

An evolutionary tree of eukaryotic organisms, constructed by the comparison of several orthologous gene sequences.

Over many generations, the genomes of organisms can change significantly, resulting in evolution. In the process called adaptation, selection for beneficial mutations can cause a species to evolve into forms better able to survive in their environment. New species are formed through the process of speciation, often caused by geographical separations that prevent populations from exchanging genes with each other. The application of genetic principles to the study of population biology and evolution is known as the "modern synthesis".

By comparing the homology between different species' genomes, it is possible to calculate the evolutionary distance between them and when they may have diverged. Genetic comparisons are generally considered a more accurate method of characterizing the relatedness between species than the comparison of phenotypic characteristics. The evolutionary distances between species can be used to form evolutionary trees; these trees represent the common descent and divergence of species over time, although they do not show the transfer of genetic material between unrelated species (known as horizontal gene transfer and most common in bacteria).

Model Organisms

Although geneticists originally studied inheritance in a wide range of organisms, researchers began to specialize in studying the genetics of a particular subset of organisms. The fact that significant research already existed for a given organism would encourage new researchers to choose it for further study, and so eventually a few model organisms became the basis for most genetics research. Common research topics in model organism genetics include the study of gene regulation and the involvement of genes in development and cancer.

Organisms were chosen, in part, for convenience—short generation times and easy genetic manipulation made some organisms popular genetics research tools. Widely used model organisms include the gut bacterium *Escherichia coli*, the plant *Arabidopsis thaliana*, baker's yeast (*Sac-*

charomyces cerevisiae), the nematode *Caenorhabditis elegans*, the common fruit fly (*Drosophila melanogaster*), and the common house mouse (*Mus musculus*).

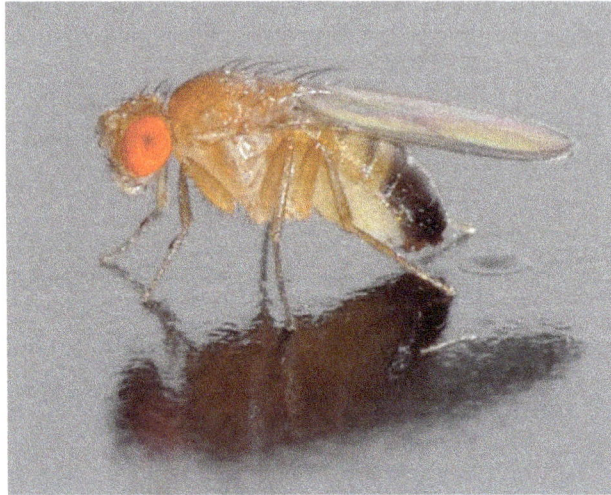

The common fruit fly (Drosophila melanogaster) is a popular model organism in genetics research.

Medicine

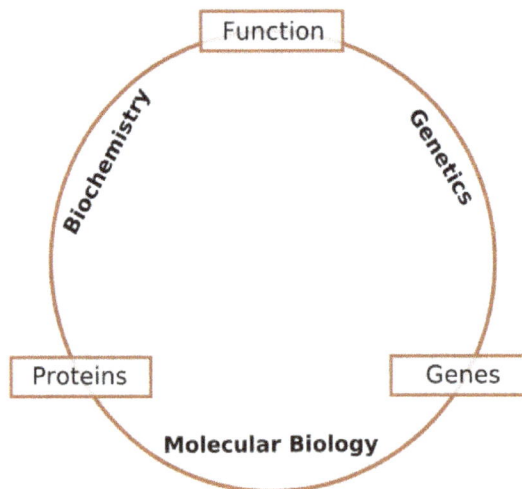

Schematic relationship between biochemistry, genetics and molecular biology.

Medical genetics seeks to understand how genetic variation relates to human health and disease. When searching for an unknown gene that may be involved in a disease, researchers commonly use genetic linkage and genetic pedigree charts to find the location on the genome associated with the disease. At the population level, researchers take advantage of Mendelian randomization to look for locations in the genome that are associated with diseases, a method especially useful for multigenic traits not clearly defined by a single gene. Once a candidate gene is found, further research is often done on the corresponding gene – the orthologous gene – in model organisms. In addition to studying genetic diseases, the increased availability of genotyping methods has led to the field of pharmacogenetics: the study of how genotype can affect drug responses.

Individuals differ in their inherited tendency to develop cancer, and cancer is a genetic disease.

The process of cancer development in the body is a combination of events. Mutations occasionally occur within cells in the body as they divide. Although these mutations will not be inherited by any offspring, they can affect the behavior of cells, sometimes causing them to grow and divide more frequently. There are biological mechanisms that attempt to stop this process; signals are given to inappropriately dividing cells that should trigger cell death, but sometimes additional mutations occur that cause cells to ignore these messages. An internal process of natural selection occurs within the body and eventually mutations accumulate within cells to promote their own growth, creating a cancerous tumor that grows and invades various tissues of the body.

Normally, a cell divides only in response to signals called growth factors and stops growing once in contact with surrounding cells and in response to growth-inhibitory signals. It usually then divides a limited number of times and dies, staying within the epithelium where it is unable to migrate to other organs. To become a cancer cell, a cell has to accumulate mutations in a number of genes (3–7) that allow it to bypass this regulation: it no longer needs growth factors to divide, it continues growing when making contact to neighbor cells, and ignores inhibitory signals, it will keep growing indefinitely and is immortal, it will escape from the epithelium and ultimately may be able to escape from the primary tumor, cross the endothelium of a blood vessel, be transported by the bloodstream and will colonize a new organ, forming deadly metastasis. Although there are some genetic predispositions in a small fraction of cancers, the major fraction is due to a set of new genetic mutations that originally appear and accumulate in one or a small number of cells that will divide to form the tumor and are not transmitted to the progeny (somatic mutations). The most frequent mutations are a loss of function of p53 protein, a tumor suppressor, or in the p53 pathway, and gain of function mutations in the ras proteins, or in other oncogenes.

Research Methods

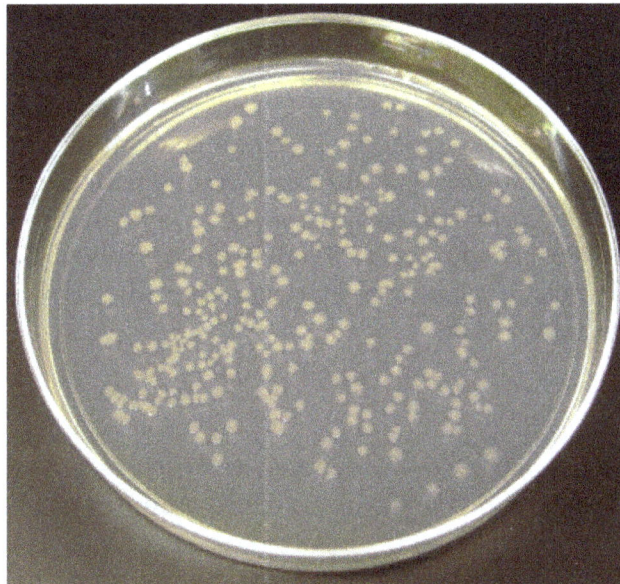

Colonies of E. coli produced by cellular cloning. A similar methodology is often used in molecular cloning.

DNA can be manipulated in the laboratory. Restriction enzymes are commonly used enzymes that cut DNA at specific sequences, producing predictable fragments of DNA. DNA fragments can be visualized through use of gel electrophoresis, which separates fragments according to their length.

The use of ligation enzymes allows DNA fragments to be connected. By binding ("ligating") fragments of DNA together from different sources, researchers can create recombinant DNA, the DNA often associated with genetically modified organisms. Recombinant DNA is commonly used in the context of plasmids: short circular DNA molecules with a few genes on them. In the process known as molecular cloning, researchers can amplify the DNA fragments by inserting plasmids into bacteria and then culturing them on plates of agar (to isolate clones of bacteria cells). ("Cloning" can also refer to the various means of creating cloned ("clonal") organisms.)

DNA can also be amplified using a procedure called the polymerase chain reaction (PCR). By using specific short sequences of DNA, PCR can isolate and exponentially amplify a targeted region of DNA. Because it can amplify from extremely small amounts of DNA, PCR is also often used to detect the presence of specific DNA sequences.

DNA Sequencing and Genomics

DNA sequencing, one of the most fundamental technologies developed to study genetics, allows researchers to determine the sequence of nucleotides in DNA fragments. The technique of chain-termination sequencing, developed in 1977 by a team led by Frederick Sanger, is still routinely used to sequence DNA fragments. Using this technology, researchers have been able to study the molecular sequences associated with many human diseases.

As sequencing has become less expensive, researchers have sequenced the genomes of many organisms, using a process called genome assembly, which utilizes computational tools to stitch together sequences from many different fragments. These technologies were used to sequence the human genome in the Human Genome Project completed in 2003. New high-throughput sequencing technologies are dramatically lowering the cost of DNA sequencing, with many researchers hoping to bring the cost of resequencing a human genome down to a thousand dollars.

Next generation sequencing (or high-throughput sequencing) came about due to the ever-increasing demand for low-cost sequencing. These sequencing technologies allow the production of potentially millions of sequences concurrently. The large amount of sequence data available has created the field of genomics, research that uses computational tools to search for and analyze patterns in the full genomes of organisms. Genomics can also be considered a subfield of bioinformatics, which uses computational approaches to analyze large sets of biological data. A common problem to these fields of research is how to manage and share data that deals with human subject and personally identifiable information.

Society and Culture

On 19 March 2015, a leading group of biologists urged a worldwide ban on clinical use of methods, particularly the use of CRISPR and zinc finger, to edit the human genome in a way that can be inherited. In April 2015, Chinese researchers reported results of basic research to edit the DNA of non-viable human embryos using CRISPR.

References

- Griffiths, Anthony J. F.; Miller, Jeffrey H.; Suzuki, David T.; Lewontin, Richard C.; Gelbart, eds. (2000). "Genetics and the Organism: Introduction". An Introduction to Genetic Analysis (7th ed.). New York: W. H. Freeman. ISBN 0-7167-3520-2.

- Cell and Molecular Biology", Pragya Khanna. I. K. International Pvt Ltd, 2008. p. 221. ISBN 81-89866-59-1, ISBN 978-81-89866-59-4

- Judson, Horace (1979). The Eighth Day of Creation: Makers of the Revolution in Biology. Cold Spring Harbor Laboratory Press. pp. 51–169. ISBN 0-87969-477-7.

- Griffiths, Anthony J. F.; Miller, Jeffrey H.; Suzuki, David T.; Lewontin, Richard C.; Gelbart, eds. (2000). "Patterns of Inheritance: Introduction". An Introduction to Genetic Analysis (7th ed.). New York: W. H. Freeman. ISBN 0-7167-3520-2.

- Griffiths, Anthony J. F.; Miller, Jeffrey H.; Suzuki, David T.; Lewontin, Richard C.; Gelbart, eds. (2000). "Mendel's experiments". An Introduction to Genetic Analysis (7th ed.). New York: W. H. Freeman. ISBN 0-7167-3520-2.

- Wade, Nicholas (19 March 2015). "Scientists Seek Ban on Method of Editing the Human Genome". New York Times. Retrieved 20 March 2015.

- Lanphier, Edward; Urnov, Fyodor; Haecker, Sarah Ehlen; Werner, Michael; Smolenski, Joanna (26 March 2015). "Don't edit the human germ line". Nature. 519: 410–411. Bibcode:2015Natur.519..410L. doi:10.1038/519410a. PMID 25810189. Retrieved 20 March 2015.

- Kolata, Gina (23 April 2015). "Chinese Scientists Edit Genes of Human Embryos, Raising Concerns". New York Times. Retrieved 24 April 2015.

- Liang, Puping; et al. (18 April 2015). "CRISPR/Cas9-mediated gene editing in human tripronuclear zygotes". Protein & Cell. 6: 363–72. doi:10.1007/s13238-015-0153-5. PMC 4417674. PMID 25894090. Retrieved 24 April 2015

Key Components of Genetics

To understand the complex science of genetics, it is of utmost importance that one grasps its key building blocks. The major components of genetics like gene, heredity, DNA, RNA, chromosome, DNA sequencing, and nucleotide have been discussed in the chapter. This enables the reader to form a comprehensive perception of genetics.

Gene

A gene is a locus (or region) of DNA which is made up of nucleotides and is the molecular unit of heredity.:Glossary The transmission of genes to an organism's offspring is the basis of the inheritance of phenotypic traits. Most biological traits are under the influence of polygenes (many different genes) as well as the gene–environment interactions. Some genetic traits are instantly visible, such as eye colour or number of limbs, and some are not, such as blood type, risk for specific diseases, or the thousands of basic biochemical processes that comprise life. In July 2016, scientists reported identifying a set of 355 genes from the Last Universal Common Ancestor (LUCA) of all organisms living on Earth.

Genes can acquire mutations in their sequence, leading to different variants, known as alleles, in the population. These alleles encode slightly different versions of a protein, which cause different phenotype traits. Colloquial usage of the term "having a gene" (e.g., "good genes," "hair colour gene") typically refers to having a different allele of the gene. Genes evolve due to natural selection or survival of the fittest of the alleles.

The concept of a gene continues to be refined as new phenomena are discovered. For example, regulatory regions of a gene can be far removed from its coding regions, and coding regions can be split into several exons. Some viruses store their genome in RNA instead of DNA and some gene products are functional non-coding RNAs. Therefore, a broad, modern working definition of a gene is any discrete locus of heritable, genomic sequence which affect an organism's traits by being expressed as a functional product or by regulation of gene expression.

History

Gregor Mendel

Discovery of Discrete Inherited Units

The existence of discrete inheritable units was first suggested by Gregor Mendel (1822–1884). From 1857 to 1864, he studied inheritance patterns in 8000 common edible pea plants, tracking distinct traits from parent to offspring. He described these mathematically as 2^n combinations where n is the number of differing characteristics in the original peas. Although he did not use the term *gene*, he explained his results in terms of discrete inherited units that give rise to observable physical characteristics. This description prefigured the distinction between genotype (the genetic material of an organism) and phenotype (the visible traits of that organism). Mendel was also the first to demonstrate independent assortment, the distinction between dominant and recessive traits, the distinction between a heterozygote and homozygote, and the phenomenon of discontinuous inheritance.

Prior to Mendel's work, the dominant theory of heredity was one of blending inheritance, which suggested that each parent contributed fluids to the fertilisation process and that the traits of the parents blended and mixed to produce the offspring. Charles Darwin developed a theory of inheritance he termed pangenesis, from Greek pan ("all, whole") and genesis ("birth") / genos ("origin"). Darwin used the term *gemmule* to describe hypothetical particles that would mix during reproduction.

Mendel's work went largely unnoticed after its first publication in 1866, but was rediscovered in the late 19th-century by Hugo de Vries, Carl Correns, and Erich von Tschermak, who (claimed to have) reached similar conclusions in their own research. Specifically, in 1889, Hugo de Vries published his book *Intracellular Pangenesis*, in which he postulated that different characters have individual hereditary carriers and that inheritance of specific traits in organisms comes in particles. De Vries called these units "pangenes" (*Pangens* in German), after Darwin's 1868 pangenesis theory.

Sixteen years later, in 1905, the word *genetics* was first used by William Bateson, while Eduard Strasburger, amongst others, still used the term pangene for the fundamental physical and functional unit of heredity. In 1909 the Danish botanist Wilhelm Johannsen shortened the name to "gene".

Discovery of DNA

Advances in understanding genes and inheritance continued throughout the 20th century. Deoxyribonucleic acid (DNA) was shown to be the molecular repository of genetic information by experiments in the 1940s to 1950s. The structure of DNA was studied by Rosalind Franklin and Maurice Wilkins using X-ray crystallography, which led James D. Watson and Francis Crick to publish a model of the double-stranded DNA molecule whose paired nucleotide bases indicated a compelling hypothesis for the mechanism of genetic replication. Collectively, this body of research established the central dogma of molecular biology, which states that proteins are translated from RNA, which is transcribed from DNA. This dogma has since been shown to have exceptions, such as reverse transcription in retroviruses. The modern study of genetics at the level of DNA is known as molecular genetics.

In 1972, Walter Fiers and his team at the University of Ghent were the first to determine the sequence of a gene: the gene for Bacteriophage MS2 coat protein. The subsequent development of chain-termination DNA sequencing in 1977 by Frederick Sanger improved the efficiency of sequencing and turned it into a routine laboratory tool. An automated version of the Sanger method was used in early phases of the Human Genome Project.

Modern Evolutionary Synthesis

The theories developed in the 1930s and 1940s to integrate molecular genetics with Darwinian evolution are called the modern evolutionary synthesis, a term introduced by Julian Huxley. Evolutionary biologists subsequently refined this concept, such as George C. Williams' gene-centric view of evolution. He proposed an evolutionary concept of the gene as a unit of natural selection with the definition: "that which segregates and recombines with appreciable frequency." In this view, the molecular gene *transcribes* as a unit, and the evolutionary gene *inherits* as a unit. Related ideas emphasizing the centrality of genes in evolution were popularized by Richard Dawkins.

Molecular Basis

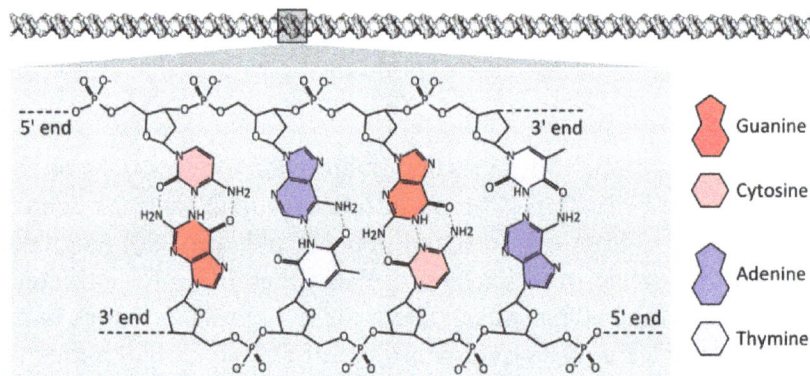

The chemical structure of a four base pair fragment of a DNA double helix. The sugar-phosphate backbone chains run in opposite directions with the bases pointing inwards, base-pairing A to T and C to G with hydrogen bonds.

DNA

The vast majority of living organisms encode their genes in long strands of DNA (deoxyribonucleic acid). DNA consists of a chain made from four types of nucleotide subunits, each composed of: a five-carbon sugar (2'-deoxyribose), a phosphate group, and one of the four bases adenine, cytosine, guanine, and thymine.

Two chains of DNA twist around each other to form a DNA double helix with the phosphate-sugar backbone spiralling around the outside, and the bases pointing inwards with adenine base pairing to thymine and guanine to cytosine. The specificity of base pairing occurs because adenine and thymine align to form two hydrogen bonds, whereas cytosine and guanine form three hydrogen bonds. The two strands in a double helix must therefore be complementary, with their sequence of bases matching such that the adenines of one strand are paired with the thymines of the other strand, and so on.

Due to the chemical composition of the pentose residues of the bases, DNA strands have directionality. One end of a DNA polymer contains an exposed hydroxyl group on the deoxyribose; this is known as the 3' end of the molecule. The other end contains an exposed phosphate group; this is the 5' end. The two strands of a double-helix run in opposite directions. Nucleic acid synthesis, including DNA replication and transcription occurs in the 5'→3' direction, because new nucleotides are added via a dehydration reaction that uses the exposed 3' hydroxyl as a nucleophile.

The expression of genes encoded in DNA begins by transcribing the gene into RNA, a second type of nucleic acid that is very similar to DNA, but whose monomers contain the sugar ribose rather than deoxyribose. RNA also contains the base uracil in place of thymine. RNA molecules are less stable than DNA and are typically single-stranded. Genes that encode proteins are composed of a series of three-nucleotide sequences called codons, which serve as the "words" in the genetic "language". The genetic code specifies the correspondence during protein translation between codons and amino acids. The genetic code is nearly the same for all known organisms.

Chromosomes

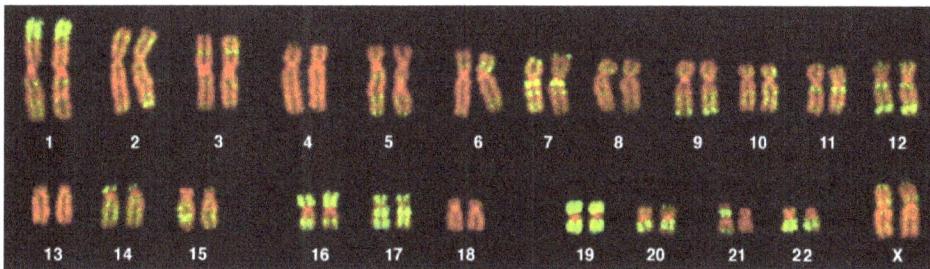

Fluorescent microscopy image of a human female karyotype, showing 23 pairs of chromosomes . The DNA is stained red, with regions rich in housekeeping genes further stained in green. The largest chromosomes are around 10 times the size of the smallest.

The total complement of genes in an organism or cell is known as its genome, which may be stored on one or more chromosomes. A chromosome consists of a single, very long DNA helix on which thousands of genes are encoded. The region of the chromosome at which a particular gene is located is called its locus. Each locus contains one allele of a gene; however, members of a population may have different alleles at the locus, each with a slightly different gene sequence.

The majority of eukaryotic genes are stored on a set of large, linear chromosomes. The chromosomes are packed within the nucleus in complex with storage proteins called histones to form a unit called a nucleosome. DNA packaged and condensed in this way is called chromatin. The manner in which DNA is stored on the histones, as well as chemical modifications of the histone itself, regulate whether a particular region of DNA is accessible for gene expression. In addition to genes, eukaryotic chromosomes contain sequences involved in ensuring that the DNA is copied without degradation of end regions and sorted into daughter cells during cell division: replication origins, telomeres and the centromere. Replication origins are the sequence regions where DNA replication is initiated to make two copies of the chromosome. Telomeres are long stretches of repetitive sequence that cap the ends of the linear chromosomes and prevent degradation of coding and regulatory regions during DNA replication. The length of the telomeres decreases each time the genome is replicated and has been implicated in the aging process. The centromere is required for binding spindle fibres to separate sister chromatids into daughter cells during cell division.

Prokaryotes (bacteria and archaea) typically store their genomes on a single large, circular chromosome. Similarly, some eukaryotic organelles contain a remnant circular chromosome with a small number of genes. Prokaryotes sometimes supplement their chromosome with additional small circles of DNA called plasmids, which usually encode only a few genes and are transferable between individuals. For example, the genes for antibiotic resistance are usually encoded on bacterial plasmids and can be passed between individual cells, even those of different species, via horizontal gene transfer.

Whereas the chromosomes of prokaryotes are relatively gene-dense, those of eukaryotes often contain regions of DNA that serve no obvious function. Simple single-celled eukaryotes have relatively small amounts of such DNA, whereas the genomes of complex multicellular organisms, including humans, contain an absolute majority of DNA without an identified function. This DNA has often been referred to as "junk DNA". However, more recent analyses suggest that, although protein-coding DNA makes up barely 2% of the human genome, about 80% of the bases in the genome may be expressed, so the term "junk DNA" may be a misnomer.

Structure and Function

The structure of a gene consists of many elements of which the actual protein coding sequence is often only a small part. These include DNA regions that are not transcribed as well as untranslated regions of the RNA.

Firstly, flanking the open reading frame, all genes contain a regulatory sequence that is required for their expression. In order to be expressed, genes require a promoter sequence. The promoter is recognized and bound by transcription factors and RNA polymerase to initiate transcription. A gene can have more than one promoter, resulting in messenger RNAs (mRNA) that differ in how far they extend in the 5' end. Promoter regions have a consensus sequence, however highly transcribed genes have "strong" promoter sequences that bind the transcription machinery well, whereas others have "weak" promoters that bind poorly and initiate transcription less frequently. Eukaryotic promoter regions are much more complex and difficult to identify than prokaryotic promoters.

Additionally, genes can have regulatory regions many kilobases upstream or downstream of the open reading frame. These act by binding to transcription factors which then cause the DNA to loop so that the regulatory sequence (and bound transcription factor) become close to the RNA polymerase binding site. For example, enhancers increase transcription by binding an activator protein which then helps to recruit the RNA polymerase to the promoter; conversely silencers bind repressor proteins and make the DNA less available for RNA polymerase.

The transcribed pre-mRNA contains untranslated regions at both ends which contain a ribosome binding site, terminator and start and stop codons. In addition, most eukaryotic open reading frames contain untranslated introns which are removed before the exons are translated. The sequences at the ends of the introns, dictate the splice sites to generate the final mature mRNA which encodes the protein or RNA product.

Many prokaryotic genes are organized into operons, with multiple protein-coding sequences that are transcribed as a unit. The products of operon genes typically have related functions and are involved in the same regulatory network.

Functional Definitions

Defining exactly what section of a DNA sequence comprises a gene is difficult. Regulatory regions of a gene such as enhancers do not necessarily have to be close to the coding sequence on the linear molecule because the intervening DNA can be looped out to bring the gene and its regulatory region into proximity. Similarly, a gene's introns can be much larger than its exons. Regulatory regions can even be on entirely different chromosomes and operate *in trans* to allow regulatory regions on one chromosome to come in contact with target genes on another chromosome.

Early work in molecular genetics suggested the model that one gene makes one protein. This model has been refined since the discovery of genes that can encode multiple proteins by alternative splicing and coding sequences split in short section across the genome whose mRNAs are concatenated by trans-splicing.

A broad operational definition is sometimes used to encompass the complexity of these diverse phenomena, where a gene is defined as a union of genomic sequences encoding a coherent set of potentially overlapping functional products. This definition categorizes genes by their functional products (proteins or RNA) rather than their specific DNA loci, with regulatory elements classified as *gene-associated* regions.

Gene Expression

In all organisms, two steps are required to read the information encoded in a gene's DNA and produce the protein it specifies. First, the gene's DNA is *transcribed* to messenger RNA (mRNA). Second, that mRNA is *translated* to protein. RNA-coding genes must still go through the first step, but are not translated into protein. The process of producing a biologically functional molecule of either RNA or protein is called gene expression, and the resulting molecule is called a gene product.

Genetic Code

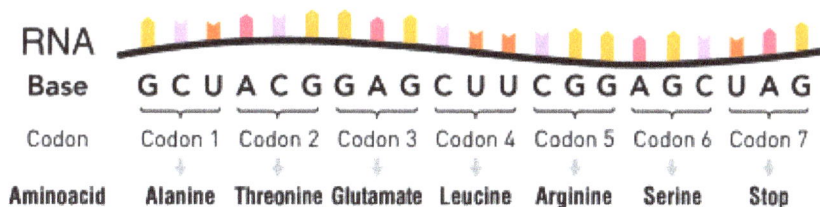

Schematic of a single-stranded RNA molecule illustrating a series of three-base codons. Each three-nucleotide codon corresponds to an amino acid when translated to protein

The nucleotide sequence of a gene's DNA specifies the amino acid sequence of a protein through the genetic code. Sets of three nucleotides, known as codons, each correspond to a specific amino acid. Additionally, a "start codon", and three "stop codons" indicate the beginning and end of the protein coding region. There are 64 possible codons (four possible nucleotides at each of three positions, hence 4^3 possible codons) and only 20 standard amino acids; hence the code is redundant and multiple codons can specify the same amino acid. The correspondence between codons and amino acids is nearly universal among all known living organisms.

Transcription produces a single-stranded RNA molecule known as messenger RNA, whose nucleotide sequence is complementary to the DNA from which it was transcribed. The mRNA acts as an intermediate between the DNA gene and its final protein product. The gene's DNA is used as a template to generate a complementary mRNA. The mRNA matches the sequence of the gene's DNA coding strand because it is synthesised as the complement of the template strand. Transcription is performed by an enzyme called an RNA polymerase, which reads the template strand in the 3' to 5' direction and synthesizes the RNA from 5' to 3'. To initiate transcription, the polymerase first recognizes and binds a promoter region of the gene. Thus, a major mechanism of gene regulation is the blocking or sequestering the promoter region, either by tight binding by repressor molecules that physically block the polymerase, or by organizing the DNA so that the promoter region is not accessible.

In prokaryotes, transcription occurs in the cytoplasm; for very long transcripts, translation may begin at the 5' end of the RNA while the 3' end is still being transcribed. In eukaryotes, transcription occurs in the nucleus, where the cell's DNA is stored. The RNA molecule produced by the polymerase is known as the primary transcript and undergoes post-transcriptional modifications before being exported to the cytoplasm for translation. One of the modifications performed is the splicing of introns which are sequences in the transcribed region that do not encode protein. Alternative splicing mechanisms can result in mature transcripts from the same gene having different sequences and thus coding for different proteins. This is a major form of regulation in eukaryotic cells and also occurs in some prokaryotes.

Translation

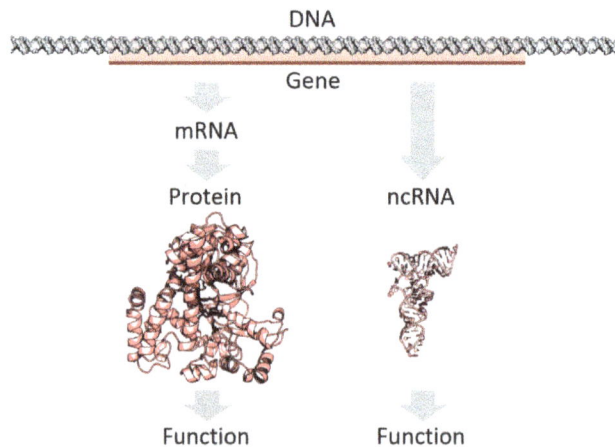

Protein coding genes are transcribed to an mRNA intermediate, then translated to a functional protein. RNA-coding genes are transcribed to a functional non-coding RNA. (PDB: 3BSE, 1OBB, 3TRA)

Translation is the process by which a mature mRNA molecule is used as a template for synthesizing a new protein. Translation is carried out by ribosomes, large complexes of RNA and protein responsible for carrying out the chemical reactions to add new amino acids to a growing polypeptide chain by the formation of peptide bonds. The genetic code is read three nucleotides at a time, in units called codons, via interactions with specialized RNA molecules called transfer RNA (tRNA). Each tRNA has three unpaired bases known as the anticodon that are complementary to the codon it reads on the mRNA. The tRNA is also covalently attached to the amino acid specified by the complementary codon. When the tRNA binds to its complementary codon in an mRNA strand, the ribosome attaches its amino acid cargo to the new polypeptide chain, which is synthesized from amino terminus to carboxyl terminus. During and after synthesis, most new proteins must folds to their active three-dimensional structure before they can carry out their cellular functions.

Regulation

Genes are regulated so that they are expressed only when the product is needed, since expression draws on limited resources. A cell regulates its gene expression depending on its external environment (e.g. available nutrients, temperature and other stresses), its internal environment (e.g. cell division cycle, metabolism, infection status), and its specific role if in a multicellular organism. Gene expression can be regulated at any step: from transcriptional initiation, to RNA processing, to post-translational modification of the protein. The regulation of lactose metabolism genes in *E. coli* (*lac* operon) was the first such mechanism to be described in 1961.

RNA Genes

A typical protein-coding gene is first copied into RNA as an intermediate in the manufacture of the final protein product. In other cases, the RNA molecules are the actual functional products, as in the synthesis of ribosomal RNA and transfer RNA. Some RNAs known as ribozymes are capable of enzymatic function, and microRNA has a regulatory role. The DNA sequences from which such RNAs are transcribed are known as non-coding RNA genes.

Some viruses store their entire genomes in the form of RNA, and contain no DNA at all. Because they use RNA to store genes, their cellular hosts may synthesize their proteins as soon as they are infected and without the delay in waiting for transcription. On the other hand, RNA retroviruses, such as HIV, require the reverse transcription of their genome from RNA into DNA before their proteins can be synthesized. RNA-mediated epigenetic inheritance has also been observed in plants and very rarely in animals.

Inheritance

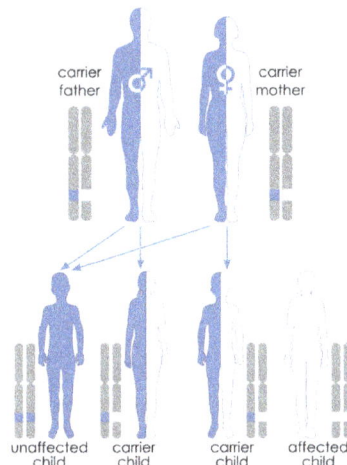

Inheritance of a gene that has two different alleles (blue and white). The gene is located on an autosomal chromosome. The blue allele is recessive to the white allele. The probability of each outcome in the children's generation is one quarter, or 25 percent.

Organisms inherit their genes from their parents. Asexual organisms simply inherit a complete copy of their parent's genome. Sexual organisms have two copies of each chromosome because they inherit one complete set from each parent.

Mendelian Inheritance

According to Mendelian inheritance, variations in an organism's phenotype (observable physical and behavioral characteristics) are due in part to variations in its genotype (particular set of genes). Each gene specifies a particular trait with different sequence of a gene (alleles) giving rise to different phenotypes. Most eukaryotic organisms (such as the pea plants Mendel worked on) have two alleles for each trait, one inherited from each parent.

Alleles at a locus may be dominant or recessive; dominant alleles give rise to their corresponding phenotypes when paired with any other allele for the same trait, whereas recessive alleles give rise to their corresponding phenotype only when paired with another copy of the same allele. For example, if the allele specifying tall stems in pea plants is dominant over the allele specifying short stems, then pea plants that inherit one tall allele from one parent and one short allele from the other parent will also have tall stems. Mendel's work demonstrated that alleles assort independently in the production of gametes, or germ cells, ensuring variation in the next generation. Although Mendelian inheritance remains a good model for many traits determined by single genes (including a number of well-known genetic disorders) it does not include the physical processes of DNA replication and cell division.

DNA Replication and Cell Division

The growth, development, and reproduction of organisms relies on cell division, or the process by which a single cell divides into two usually identical daughter cells. This requires first making a duplicate copy of every gene in the genome in a process called DNA replication. The copies are made by specialized enzymes known as DNA polymerases, which "read" one strand of the double-helical DNA, known as the template strand, and synthesize a new complementary strand. Because the DNA double helix is held together by base pairing, the sequence of one strand completely specifies the sequence of its complement; hence only one strand needs to be read by the enzyme to produce a faithful copy. The process of DNA replication is semiconservative; that is, the copy of the genome inherited by each daughter cell contains one original and one newly synthesized strand of DNA.

After DNA replication is complete, the cell must physically separate the two copies of the genome and divide into two distinct membrane-bound cells. In prokaryotes (bacteria and archaea) this usually occurs via a relatively simple process called binary fission, in which each circular genome attaches to the cell membrane and is separated into the daughter cells as the membrane invaginates to split the cytoplasm into two membrane-bound portions. Binary fission is extremely fast compared to the rates of cell division in eukaryotes. Eukaryotic cell division is a more complex process known as the cell cycle; DNA replication occurs during a phase of this cycle known as S phase, whereas the process of segregating chromosomes and splitting the cytoplasm occurs during M phase.

Molecular Inheritance

The duplication and transmission of genetic material from one generation of cells to the next is the basis for molecular inheritance, and the link between the classical and molecular pictures of genes. Organisms inherit the characteristics of their parents because the cells of the offspring contain copies of the genes in their parents' cells. In asexually reproducing organisms, the offspring will be a genetic copy or clone of the parent organism. In sexually reproducing organisms, a specialized form of cell division called meiosis produces cells called gametes or germ cells that are haploid, or contain only one copy of each gene. The gametes produced by females are called eggs or ova, and those produced by males are called sperm. Two gametes fuse to form a diploid fertilized egg, a single cell that has two sets of genes, with one copy of each gene from the mother and one from the father.

During the process of meiotic cell division, an event called genetic recombination or *crossing-over* can sometimes occur, in which a length of DNA on one chromatid is swapped with a length of DNA on the corresponding sister chromatid. This has no effect if the alleles on the chromatids are the same, but results in reassortment of otherwise linked alleles if they are different. The Mendelian principle of independent assortment asserts that each of a parent's two genes for each trait will sort independently into gametes; which allele an organism inherits for one trait is unrelated to which allele it inherits for another trait. This is in fact only true for genes that do not reside on the same chromosome, or are located very far from one another on the same chromosome. The closer two genes lie on the same chromosome, the more closely they will be associated in gametes and the more often they will appear together; genes that are very close are essentially never separated because it is extremely unlikely that a crossover point will occur between them. This is known as genetic linkage.

Molecular Evolution

Mutation

DNA replication is for the most part extremely accurate, however errors (mutations) do occur. The error rate in eukaryotic cells can be as low as 10^{-8} per nucleotide per replication, whereas for some RNA viruses it can be as high as 10^{-3}. This means that each generation, each human genome accumulates 1–2 new mutations. Small mutations can be caused by DNA replication and the aftermath of DNA damage and include point mutations in which a single base is altered and frameshift mutations in which a single base is inserted or deleted. Either of these mutations can change the gene by missense (change a codon to encode a different amino acid) or nonsense (a premature stop codon). Larger mutations can be caused by errors in recombination to cause chromosomal abnormalities including the duplication, deletion, rearrangement or inversion of large sections of a chromosome. Additionally, the DNA repair mechanisms that normally revert mutations can introduce errors when repairing the physical damage to the molecule is more important than restoring an exact copy, for example when repairing double-strand breaks.

When multiple different alleles for a gene are present in a species's population it is called polymorphic. Most different alleles are functionally equivalent, however some alleles can give rise to different phenotypic traits. A gene's most common allele is called the wild type, and rare alleles are called mutants. The genetic variation in relative frequencies of different alleles in a population is due to both natural selection and genetic drift. The wild-type allele is not necessarily the ancestor of less common alleles, nor is it necessarily fitter.

Most mutations within genes are neutral, having no effect on the organism's phenotype (silent mutations). Some mutations do not change the amino acid sequence because multiple codons encode the same amino acid (synonymous mutations). Other mutations can be neutral if they lead to amino acid sequence changes, but the protein still functions similarly with the new amino acid (e.g. conservative mutations). Many mutations, however, are deleterious or even lethal, and are removed from populations by natural selection. Genetic disorders are the result of deleterious mutations and can be due to spontaneous mutation in the affected individual, or can be inherited. Finally, a small fraction of mutations are beneficial, improving the organism's fitness and are extremely important for evolution, since their directional selection leads to adaptive evolution.

Sequence Homology

Genes with a most recent common ancestor, and thus a shared evolutionary ancestry, are known as homologs. These genes appear either from gene duplication within an organism's genome, where they are known as paralogous genes, or are the result of divergence of the genes after a speciation event, where they are known as orthologous genes, and often perform the same or similar functions in related organisms. It is often assumed that the functions of orthologous genes are more similar than those of paralogous genes, although the difference is minimal.

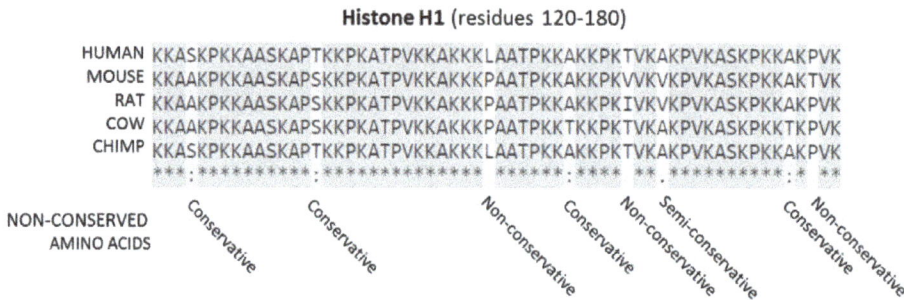

Histone H1 (residues 120-180)

HUMAN KKASKPKKAASKAPTKKPKATPVKKAKKKLAATPKKAKKPKTVKAKPVKASKPKKAKPVK
MOUSE KKAAKPKKAASKAPSKKPKATPVKKAKKKPAATPKKAKKPKVVKVKPVKASKPKKAKTVK
RAT KKAAKPKKAASKAPSKKPKATPVKKAKKKPAATPKKAKKPKIVKVKPVKASKPKKAKPVK
COW KKAAKPKKAASKAPSKKPKATPVKKAKKKPAATPKKTKKPKTVKAKPKKTKPVK
CHIMP KKASKPKKAASKAPTKKPKATPVKKAKKKLAATPKKAKKPKTVKAKPVKASKPKKAKPVK

A sequence alignment, produced by ClustalO, of mammalian histone proteins

The relationship between genes can be measured by comparing the sequence alignment of their DNA. The degree of sequence similarity between homologous genes is called conserved sequence. Most changes to a gene's sequence do not affect its function and so genes accumulate mutations over time by neutral molecular evolution. Additionally, any selection on a gene will cause its sequence to diverge at a different rate. Genes under stabilizing selection are constrained and so change more slowly whereas genes under directional selection change sequence more rapidly. The sequence differences between genes can be used for phylogenetic analyses to study how those genes have evolved and how the organisms they come from are related.

Origins of New Genes

The most common source of new genes in eukaryotic lineages is gene duplication, which creates copy number variation of an existing gene in the genome. The resulting genes (paralogs) may then diverge in sequence and in function. Sets of genes formed in this way comprise a gene family. Gene duplications and losses within a family are common and represent a major source of evolutionary biodiversity. Sometimes, gene duplication may result in a nonfunctional copy of a gene, or a functional copy may be subject to mutations that result in loss of function; such nonfunctional genes are called pseudogenes.

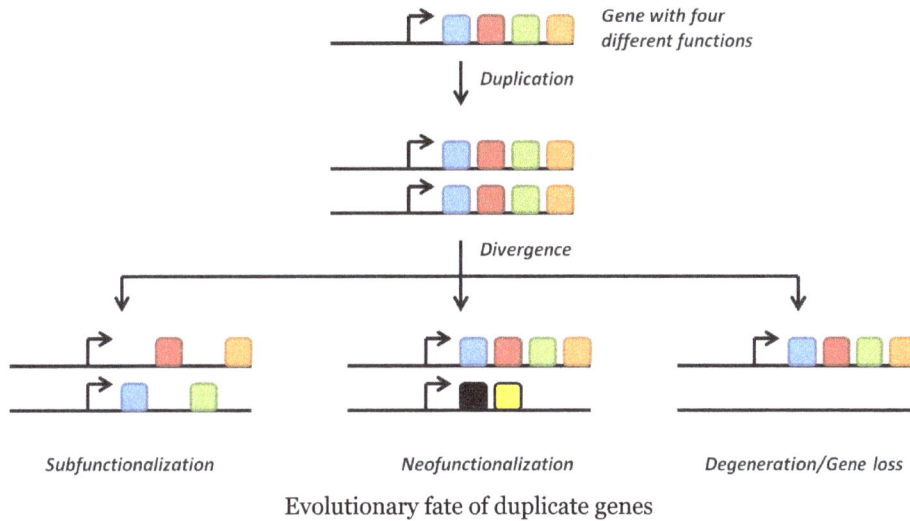

Evolutionary fate of duplicate genes

De novo or "orphan" genes, whose sequence shows no similarity to existing genes, are extremely rare. Estimates of the number of de novo genes in the human genome range from 18 to 60. Such genes are typically shorter and simpler in structure than most eukaryotic genes, with few if any introns. Two primary sources of orphan protein-coding genes are gene duplication followed by extremely rapid sequence change, such that the original relationship is undetectable by sequence comparisons, and formation through mutation of "cryptic" transcription start sites that introduce a new open reading frame in a region of the genome that did not previously code for a protein.

Horizontal gene transfer refers to the transfer of genetic material through a mechanism other than reproduction. This mechanism is a common source of new genes in prokaryotes, sometimes thought to contribute more to genetic variation than gene duplication. It is a common means of spreading antibiotic resistance, virulence, and adaptive metabolic functions. Although horizontal gene transfer is rare in eukaryotes, likely examples have been identified of protist and alga genomes containing genes of bacterial origin.

Genome

The genome is the total genetic material of an organism and includes both the genes and non-coding sequences.

Number of Genes

The genome size, and the number of genes it encodes varies widely between organisms. The smallest genomes occur in viruses (which can have as few as 2 protein-coding genes), and viroids (which act as a single non-coding RNA gene). Conversely, plants can have extremely large genomes, with rice containing >46,000 protein-coding genes. The total number of protein-coding genes (the Earth's proteome) is estimated to be 5 million sequences.

Representative genome sizes for plants (green), vertebrates (blue), invertebrates (red), fungus (yellow), bacteria (purple), and viruses (grey). An inset on the right shows the smaller genomes expanded 100-fold.

Although the number of base-pairs of DNA in the human genome has been known since the 1960s, the estimated number of genes has changed over time as definitions of genes, and methods of detecting them have been refined. Initial theoretical predictions of the number of human genes were as high as 2,000,000. Early experimental measures indicated there to be 50,000–100,000 *transcribed* genes (expressed sequence tags). Subsequently, the sequencing in the Human Genome Project indicated that many of these transcripts were alternative variants of the same genes, and the total number of protein-coding genes was revised down to ~20,000 with 13 genes encoded on the mitochondrial genome. Of the human genome, only 1–2% consists of protein-coding genes, with the remainder being 'noncoding' DNA such as introns, retrotransposons, and noncoding RNAs. Every organism has all his genes in all cells of his body but it is not important that every gene must function in every cell .

Essential Genes

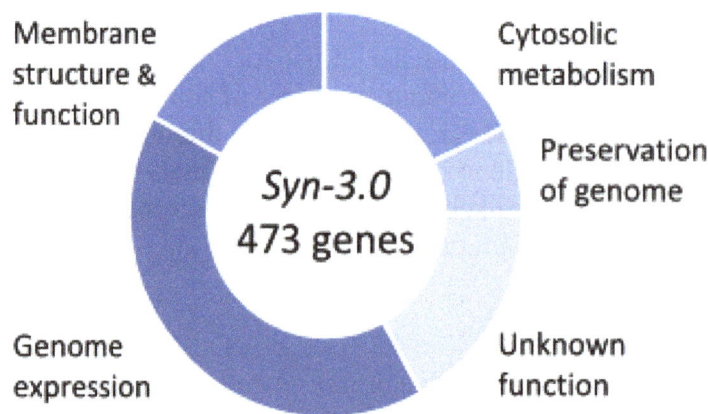

Gene functions in the minimal genome of the synthetic organism, Syn 3.

Essential genes are the set of genes thought to be critical for an organism's survival. This definition assumes the abundant availability of all relevant nutrients and the absence of environmental stress. Only a small portion of an organism's genes are essential. In bacteria, an estimated 250–400 genes are essential for *Escherichia coli* and *Bacillus subtilis*, which is less than 10% of their genes. Half of these genes are orthologs in both organisms and are largely involved in protein synthesis. In the budding yeast *Saccharomyces cerevisiae* the number of essential genes is slightly higher, at 1000 genes (~20% of their genes). Although the number is more difficult to measure in higher eukaryotes, mice and humans are estimated to have around 2000 essential genes (~10% of their genes). The synthetic organism, *Syn 3*, has a minimal genome of 473 essential genes and quasi-essential genes (necessary for fast growth), although 149 have unknown function.

Essential genes include Housekeeping genes (critical for basic cell functions) as well as genes that are expressed at different times in the organisms development or life cycle. Housekeeping genes are used as experimental controls when analysing gene expression, since they are constitutively expressed at a relatively constant level.

Genetic and Genomic Nomenclature

Gene nomenclature has been established by the HUGO Gene Nomenclature Committee (HGNC) for each known human gene in the form of an approved gene name and symbol (short-form abbreviation), which can be accessed through a database maintained by HGNC. Symbols are chosen to be unique, and each gene has only one symbol (although approved symbols sometimes change). Symbols are preferably kept consistent with other members of a gene family and with homologs in other species, particularly the mouse due to its role as a common model organism.

Genetic Engineering

Genetic engineering is the modification of an organism's genome through biotechnology. Since the 1970s, a variety of techniques have been developed to specifically add, remove and edit genes in an organism. Recently developed genome engineering techniques use engineered nuclease enzymes to create targeted DNA repair in a chromosome to either disrupt or edit a gene when the break is repaired. The related term synthetic biology is sometimes used to refer to extensive genetic engineering of an organism.

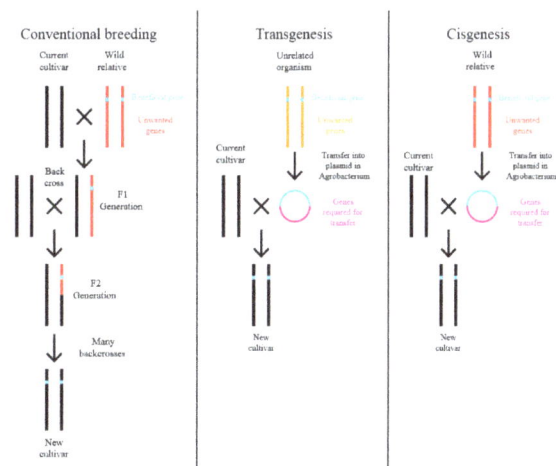

Comparison of conventional plant breeding with transgenic and cisgenic genetic modification.

Genetic engineering is now a routine research tool with model organisms. For example, genes are easily added to bacteria and lineages of knockout mice with a specific gene's function disrupted are used to investigate that gene's function. Many organisms have been genetically modified for applications in agriculture, industrial biotechnology, and medicine.

For multicellular organisms, typically the embryo is engineered which grows into the adult genetically modified organism. However, the genomes of cells in an adult organism can be edited using gene therapy techniques to treat genetic diseases.

Nucleotide

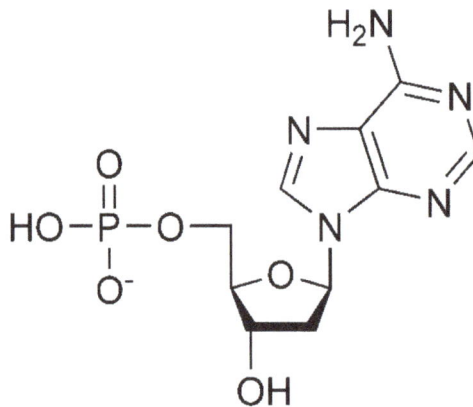

This nucleotide contains the five-carbon sugar deoxyribose, a nitrogenous base called adenine, and one phosphate group. Together, the deoxyribose and adenine make up a nucleoside (specifically, a deoxyribonucleoside) called deoxyadenosine. With the one phosphate group included, whole structure is considered a deoxyribonucleotide (a nucleotide constituent of DNA) with the name deoxyadenosine monophosphate.

Nucleotides are organic molecules that serve as the monomers, or subunits, of nucleic acids like DNA (deoxyribonucleic acid) and RNA (ribonucleic acid). The building blocks of nucleic acids, nucleotides are composed of a nitrogenous base, a five-carbon sugar (ribose or deoxyribose), and at least one phosphate group. Thus a nucleoside plus a phosphate group yields a nucleotide.

Nucleotides also function to carry packets of energy within the cell in the form of the nucleoside triphosphates (ATP, GTP, CTP and UTP), playing a central role in metabolism. In addition, nucleotides participate in cell signaling (cGMP and cAMP), and are incorporated into important cofactors of enzymatic reactions (e.g. coenzyme A, FAD, FMN, NAD, and NADP$^+$).

In experimental biochemistry, nucleotides can be radiolabeled with radionuclides to yield radionucleotides.

Structure

A nucleotide is made of a nucleobase (also termed a nitrogenous base), a five-carbon sugar (either ribose or 2-deoxyribose, depending on if it is RNA or DNA), and one or, depending on the definition, more than one phosphate groups. Authoritative chemistry sources such as the ACS Style Guide and IUPAC Gold Book clearly state that the term nucleotide refers only to a molecule containing one phosphate. However, common usage in molecular biology textbooks often extends this definition to include molecules with two or three phosphate groups. Thus, the term "nucleotide" generally refers to a nucleoside monophosphate, but a nucleoside diphosphate or nucleoside triphosphate could be considered a nucleotide as well.

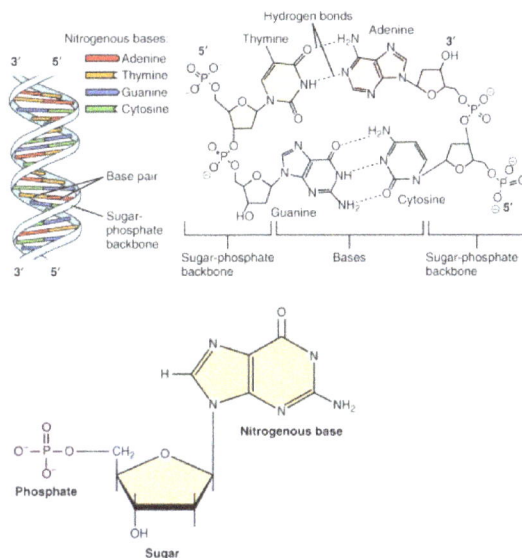

The structure of nucleotide monomers.

Without the phosphate group, the nucleobase and sugar compose a nucleoside. The phosphate groups form bonds with either the 2, 3, or 5-carbon of the sugar, with the 5-carbon site most common. Cyclic nucleotides form when the phosphate group is bound to two of the sugar's hydroxyl groups. Nucleotides contain either a purine or a pyrimidine base. Ribonucleotides are nucleotides in which the sugar is ribose. Deoxyribonucleotides are nucleotides in which the sugar is deoxyribose.

Nucleic acids are polymeric macromolecules made from nucleotide monomers. In DNA, the purine bases are adenine and guanine, while the pyrimidines are thymine and cytosine. RNA uses uracil in place of thymine. Adenine always pairs with thymine by 2 hydrogen bonds, while guanine pairs with cytosine through 3 hydrogen bonds, in each case because of the unique structures of the bases.

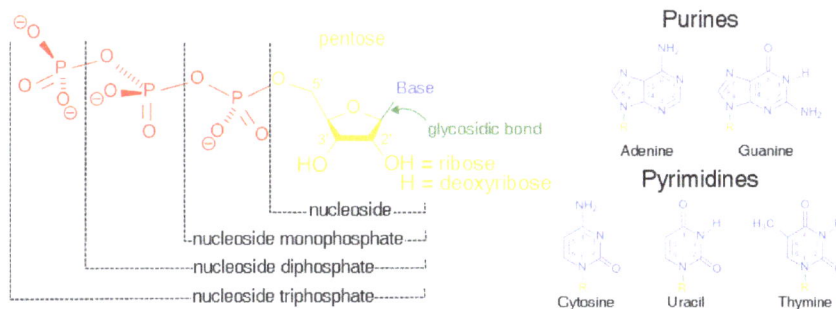

Structural elements of common nucleic acid constituents. Because they contain at least one phosphate group, the compounds marked nucleoside monophosphate, nucleoside diphosphate and nucleoside triphosphate are all nucleotides (not simply phosphate-lacking nucleosides).

Synthesis

Nucleotides can be synthesized by a variety of means both in vitro and in vivo.

In vivo, nucleotides can be synthesized de novo or recycled through salvage pathways. The components used in de novo nucleotide synthesis are derived from biosynthetic precursors of carbohydrate and amino acid metabolism, and from ammonia and carbon dioxide. The liver is the major organ of de novo synthesis of all four nucleotides. De novo synthesis of pyrimidines and purines follows two different pathways. Pyrimidines are synthesized first from aspartate and carbamoyl-phosphate in the cytoplasm to the common precursor ring structure orotic acid, onto which a phosphorylated ribosyl unit is covalently linked. Purines, however, are first synthesized from the sugar template onto which the ring synthesis occurs. For reference, the syntheses of the purine and pyrimidine nucleotides are carried out by several enzymes in the cytoplasm of the cell, not within a specific organelle. Nucleotides undergo breakdown such that useful parts can be reused in synthesis reactions to create new nucleotides.

In vitro, protecting groups may be used during laboratory production of nucleotides. A purified nucleoside is protected to create a phosphoramidite, which can then be used to obtain analogues not found in nature and/or to synthesize an oligonucleotide.

Pyrimidine Ribonucleotide Synthesis

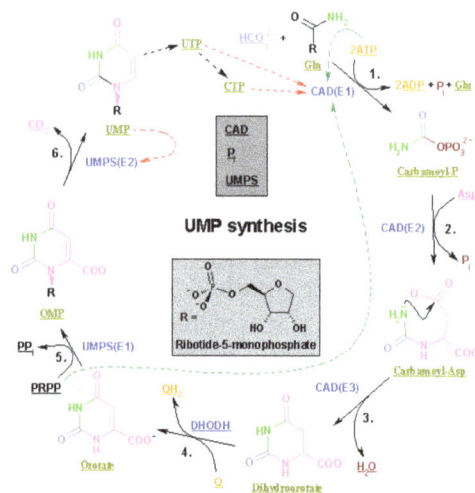

The Synthesis of UMP.

The color scheme is as follows: enzymes, coenzymes, substrate names, inorganic molecules

The synthesis of the pyrimidines CTP and UTP occurs in the cytoplasm and starts with the formation of carbamoyl phosphate from glutamine and CO_2. Next, aspartate carbamoyltransferase catalyzes a condensation reaction between aspartate and carbamoyl phosphate to form carbamoyl aspartic acid, which is cyclized into 4,5-dihydroorotic acid by dihydroorotase. The latter is converted to orotate by dihydroorotate oxidase. The net reaction is:

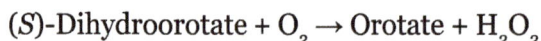

$$(S)\text{-Dihydroorotate} + O_2 \rightarrow \text{Orotate} + H_2O_2$$

Orotate is covalently linked with a phosphorylated ribosyl unit. The covalent linkage between the ribose and pyrimidine occurs at position C_1 of the ribose unit, which contains a pyrophosphate, and N_1 of the pyrimidine ring. Orotate phosphoribosyltransferase (PRPP transferase) catalyzes the net reaction yielding orotidine monophosphate (OMP):

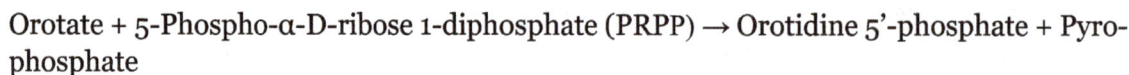

Orotate + 5-Phospho-α-D-ribose 1-diphosphate (PRPP) → Orotidine 5'-phosphate + Pyrophosphate

Orotidine 5'-monophosphate is decarboxylated by orotidine-5'-phosphate decarboxylase to form uridine monophosphate (UMP). PRPP transferase catalyzes both the ribosylation and decarboxylation reactions, forming UMP from orotic acid in the presence of PRPP. It is from UMP that other pyrimidine nucleotides are derived. UMP is phosphorylated by two kinases to uridine triphosphate (UTP) via two sequential reactions with ATP. First the diphosphate form UDP is produced, which in turn is phosphorylated to UTP. Both steps are fueled by ATP hydrolysis:

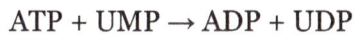

$$ATP + UMP \rightarrow ADP + UDP$$

$$UDP + ATP \rightarrow UTP + ADP$$

CTP is subsequently formed by amination of UTP by the catalytic activity of CTP synthetase. Glutamine is the NH_3 donor and the reaction is fueled by ATP hydrolysis, too:

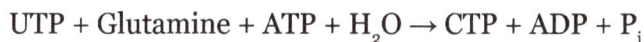

$$UTP + Glutamine + ATP + H_2O \rightarrow CTP + ADP + P_i$$

Cytidine monophosphate (CMP) is derived from cytidine triphosphate (CTP) with subsequent loss of two phosphates.

Purine Ribonucleotide Synthesis

The atoms which are used to build the purine nucleotides come from a variety of sources:

The de novo synthesis of purine nucleotides by which these precursors are incorporated into the purine ring proceeds by a 10-step pathway to the branch-point intermediate IMP, the nucleotide of the base hypoxanthine. AMP and GMP are subsequently synthesized from this intermediate via separate, two-step pathways. Thus, purine moieties are initially formed as part of the ribonucleotides rather than as free bases.

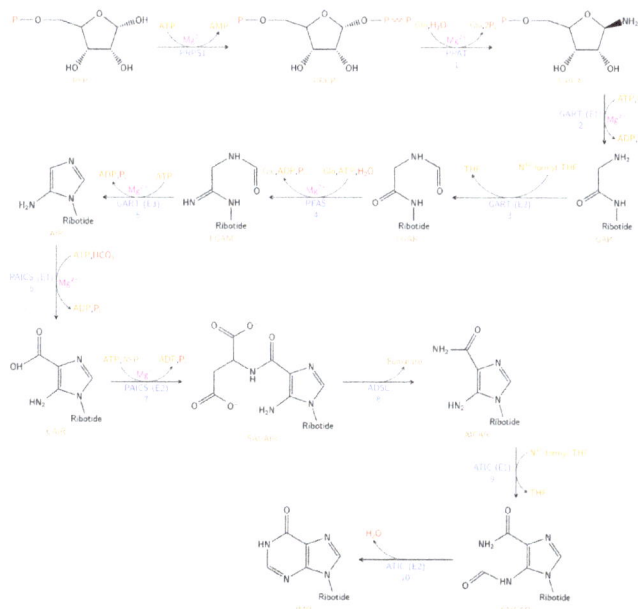

The synthesis of IMP. The color scheme is as follows: enzymes, coenzymes, substrate names, metal ions, inorganic molecules

Six enzymes take part in IMP synthesis. Three of them are multifunctional:

- GART (reactions 2, 3, and 5)

- PAICS (reactions 6, and 7)

- ATIC (reactions 9, and 10)

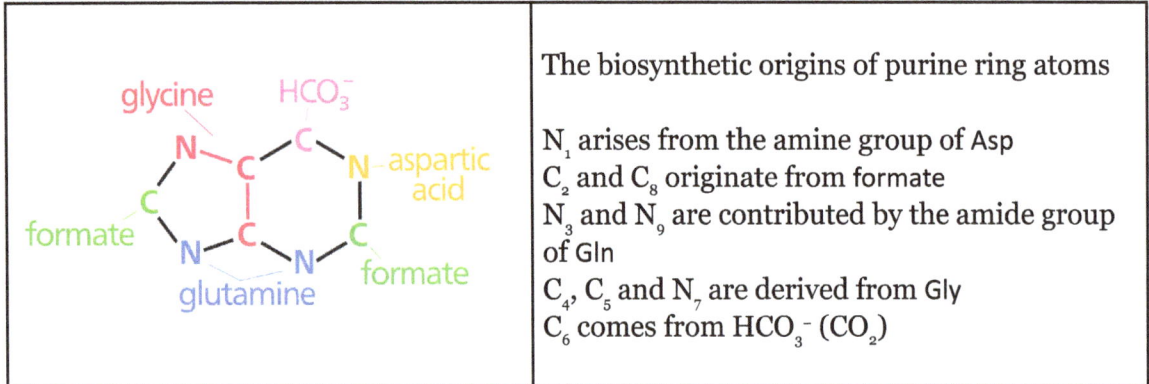

The biosynthetic origins of purine ring atoms

N_1 arises from the amine group of Asp
C_2 and C_8 originate from formate
N_3 and N_9 are contributed by the amide group of Gln
C_4, C_5 and N_7 are derived from Gly
C_6 comes from HCO_3^- (CO_2)

The pathway starts with the formation of PRPP. PRPS1 is the enzyme that activates R5P, which is formed primarily by the pentose phosphate pathway, to PRPP by reacting it with ATP. The reaction is unusual in that a pyrophosphoryl group is directly transferred from ATP to C_1 of R5P and that the product has the α configuration about C1. This reaction is also shared with the pathways for the synthesis of Trp, His, and the pyrimidine nucleotides. Being on a major metabolic crossroad and requiring much energy, this reaction is highly regulated.

In the first reaction unique to purine nucleotide biosynthesis, PPAT catalyzes the displacement of PRPP's pyrophosphate group (PP_i) by an amide nitrogen donated from either glutamine (N), glycine (N&C), aspartate (N), folic acid (C_1), or CO_2. This is the committed step in purine synthesis. The reaction occurs with the inversion of configuration about ribose C_1, thereby forming β-5-phosphorybosylamine (5-PRA) and establishing the anomeric form of the future nucleotide.

Next, a glycine is incorporated fueled by ATP hydrolysis and the carboxyl group forms an amine bond to the NH_2 previously introduced. A one-carbon unit from folic acid coenzyme N_{10}-formyl-THF is then added to the amino group of the substituted glycine followed by the closure of the imidazole ring. Next, a second NH_2 group is transferred from a glutamine to the first carbon of the glycine unit. A carboxylation of the second carbon of the glycin unit is concomittantly added. This new carbon is modified by the additional of a third NH_2 unit, this time transferred from an aspartate residue. Finally, a second one-carbon unit from formyl-THF is added to the nitrogen group and the ring covalently closed to form the common purine precursor inosine monophosphate (IMP).

Inosine monophosphate is converted to adenosine monophosphate in two steps. First, GTP hydrolysis fuels the addition of aspartate to IMP by adenylosuccinate synthase, substituting the carbonyl oxygen for a nitrogen and forming the intermediate adenylosuccinate. Fumarate is then cleaved off forming adenosine monophosphate. This step is catalyzed by adenylosuccinate lyase.

Inosine monophosphate is converted to guanosine monophosphate by the oxidation of IMP forming xanthylate, followed by the insertion of an amino group at C_2. NAD^+ is the electron acceptor in the oxidation reaction. The amide group transfer from glutamine is fueled by ATP hydrolysis.

Pyrimidine and Purine Degradation

In humans, pyrimidine rings (C, T, U) can be degraded completely to CO_2 and NH_3 (urea excretion). That having been said, purine rings (G, A) cannot. Instead they are degraded to the metabolically inert uric acid which is then excreted from the body. Uric acid is formed when GMP is split into the base guanine and ribose. Guanine is deaminated to xanthine which in turn is oxidized to uric acid. This last reaction is irreversible. Similarly, uric acid can be formed when AMP is deaminated to IMP from which the ribose unit is removed to form hypoxanthine. Hypoxanthine is oxidized to xanthine and finally to uric acid. Instead of uric acid secretion, guanine and IMP can be used for recycling purposes and nucleic acid synthesis in the presence of PRPP and aspartate (NH_3 donor).

Unnatural Base Pair (UBP)

An unnatural base pair (UBP) is a designed subunit (or nucleobase) of DNA which is created in a laboratory and does not occur in nature. In 2012, a group of American scientists led by Floyd Romesberg, a chemical biologist at the Scripps Research Institute in San Diego, California, published that his team designed an unnatural base pair (UBP). The two new artificial nucleotides or *Unnatural Base Pair* (UBP) were named d5SICS and dNaM. More technically, these artificial nucleotides bearing hydrophobic nucleobases, feature two fused aromatic rings that form a (d5SICS–dNaM) complex or base pair in DNA. In 2014 the same team from the Scripps Research Institute reported that they synthesized a stretch of circular DNA known as a plasmid containing natural T-A and C-G base pairs along with the best-performing UBP Romesberg's laboratory had designed, and inserted it into cells of the common bacterium *E. coli* that successfully replicated the unnatural base pairs through multiple generations. This is the first known example of a living organism passing along an expanded genetic code to subsequent generations. This was in part achieved by the addition of a supportive algal gene that expresses a nucleotide triphosphate transporter which efficiently imports the triphosphates of both d5SICSTP and dNaMTP into *E. coli* bacteria. Then, the natural bacterial replication pathways use them to accurately replicate the plasmid containing d5SICS–dNaM.

The successful incorporation of a third base pair is a significant breakthrough toward the goal of greatly expanding the number of amino acids which can be encoded by DNA, from the existing 20 amino acids to a theoretically possible 172, thereby expanding the potential for living organisms to produce novel proteins. The artificial strings of DNA do not encode for anything yet, but scientists speculate they could be designed to manufacture new proteins which could have industrial or pharmaceutical uses.

Length Unit

Nucleotide (abbreviated "nt") is a common unit of length for single-stranded nucleic acids, similar to how base pair is a unit of length for double-stranded nucleic acids.

Abbreviation Codes for Degenerate Bases

The IUPAC has designated the symbols for nucleotides. Apart from the five (A, G, C, T/U) bases, often degenerate bases are used especially for designing PCR primers. These nucleotide codes are listed here. Some primer sequences may also include the character "I", which codes for the non-standard nucleotide inosine. Inosine occurs in tRNAs, and will pair with adenine, cytosine, or thymine. This character does not appear in the following table however, because it does not represent a degeneracy. While inosine can serve a similar function as the degeneracy "D", it is an actual nucleotide, rather than a representation of a mix of nucleotides that covers each possible pairing needed.

Heredity

Heredity is the genetic information passing for traits from parents to their offspring, either through asexual reproduction or sexual reproduction. This is the process by which an offspring cell or organism acquires or becomes predisposed to the characteristics of its parent cell or organism. Through heredity, variations exhibited by individuals can accumulate and cause some species to evolve through the natural selection of specific phenotype traits. The study of heredity in biology is called genetics, which includes the field of epigenetics.

Overview

In humans, eye color is an example of an inherited characteristic: an individual might inherit the "brown-eye trait" from one of the parents. Inherited traits are controlled by genes and the complete set of genes within an organism's genome is called its genotype.

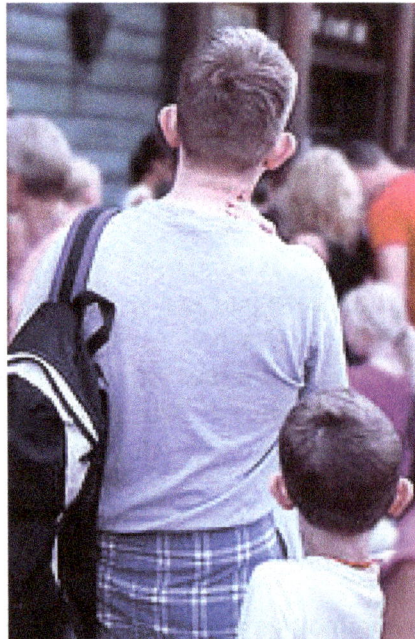

Heredity of phenotypic traits: Father and son with prominent ears and crowns.

The complete set of observable traits of the structure and behavior of an organism is called its phenotype. These traits arise from the interaction of its genotype with the environment. As a result, many aspects of an organism's phenotype are not inherited. For example, suntanned skin comes from the interaction between a person's phenotype and sunlight; thus, suntans are not passed on to people's children. However, some people tan more easily than others, due to differences in their genotype: a striking example is people with the inherited trait of albinism, who do not tan at all and are very sensitive to sunburn.

Heritable traits are known to be passed from one generation to the next via DNA, a molecule that encodes genetic information. DNA is a long polymer that incorporates four types of bases, which are interchangeable. The sequence of bases along a particular DNA molecule specifies the genetic information: this is comparable to a sequence of letters spelling out a passage of text. Before a cell divides through mitosis, the DNA is copied, so that each of the resulting two cells will inherit the DNA sequence. A portion of a DNA molecule that specifies a single functional unit is called a gene; different genes have different sequences of bases. Within cells, the long strands of DNA form condensed structures called chromosomes. Organisms inherit genetic material from their parents in the form of homologous chromosomes, containing a unique combination of DNA sequences that code for genes. The specific location of a DNA sequence within a chromosome is known as a locus. If the DNA sequence at a particular locus varies between individuals, the different forms of this sequence are called alleles. DNA sequences can change through mutations, producing new alleles. If a mutation occurs within a gene, the new allele may affect the trait that the gene controls, altering the phenotype of the organism.

However, while this simple correspondence between an allele and a trait works in some cases, most traits are more complex and are controlled by multiple interacting genes within and among organisms. Developmental biologists suggest that complex interactions in genetic networks and communication among cells can lead to heritable variations that may underlie some of the mechanics in developmental plasticity and canalization.

Recent findings have confirmed important examples of heritable changes that cannot be explained by direct agency of the DNA molecule. These phenomena are classed as epigenetic inheritance systems that are causally or independently evolving over genes. Research into modes and mechanisms of epigenetic inheritance is still in its scientific infancy, however, this area of research has attracted much recent activity as it broadens the scope of heritability and evolutionary biology in general. DNA methylation marking chromatin, self-sustaining metabolic loops, gene silencing by RNA interference, and the three dimensional conformation of proteins (such as prions) are areas where epigenetic inheritance systems have been discovered at the organismic level. Heritability may also occur at even larger scales. For example, ecological inheritance through the process of niche construction is defined by the regular and repeated activities of organisms in their environment. This generates a legacy of effect that modifies and feeds back into the selection regime of subsequent generations. Descendants inherit genes plus environmental characteristics generated by the ecological actions of ancestors. Other examples of heritability in evolution that are not under the direct control of genes include the inheritance of cultural traits, group heritability, and symbiogenesis. These examples of heritability that operate above the gene are covered broadly under the title of multilevel or hierarchical selection, which has been a subject of intense debate in the history of evolutionary science.

Relation to Theory of Evolution

When Charles Darwin proposed his theory of evolution in 1859, one of its major problems was the lack of an underlying mechanism for heredity. Darwin believed in a mix of blending inheritance and the inheritance of acquired traits (pangenesis). Blending inheritance would lead to uniformity across populations in only a few generations and then would remove variation from a population on which natural selection could act. This led to Darwin adopting some Lamarckian ideas in later editions of *On the Origin of Species* and his later biological works. Darwin's primary approach to heredity was to outline how it appeared to work (noticing that traits that were not expressed explicitly in the parent at the time of reproduction could be inherited, that certain traits could be sex-linked, etc.) rather than suggesting mechanisms.

Darwin's initial model of heredity was adopted by, and then heavily modified by, his cousin Francis Galton, who laid the framework for the biometric school of heredity. Galton found no evidence to support the aspects of Darwin's pangenesis model, which relied on acquired traits.

The inheritance of acquired traits was shown to have little basis in the 1880s when August Weismann cut the tails off many generations of mice and found that their offspring continued to develop tails.

History

Scientists in Antiquity had a variety of ideas about heredity: Theophrastus proposed that male flowers caused female flowers to ripen; Hippocrates speculated that "seeds" were produced by various body parts and transmitted to offspring at the time of conception; and Aristotle thought that male and female semen mixed at conception. Aeschylus, in 458 BC, proposed the male as the parent, with the female as a "nurse for the young life sown within her".

Ancient understandings of heredity transitioned to two debated doctrines in the 18th century. The Doctrine of Epigenesis and the Doctrine of Preformation were two distinct views of the understanding of heredity. The Doctrine of Epigenesis, originated by Aristotle, claimed that an embryo continually develops. The modifications of the parent's traits are passed off to an embryo during its lifetime. The foundation of this doctrine was based on the theory of inheritance of acquired traits. In direct opposition, the Doctrine of Preformation claimed that "like generates like" where the germ would evolve to yield offspring similar to the parents. The Preformationist view believed procreation was an act of revealing what had been created long before. However, this was disputed by the creation of the cell theory in the 19th century, where the fundamental unit of life is the cell, and not some preformed parts of an organism. Various hereditary mechanisms, including blending inheritance were also envisaged without being properly tested or quantified, and were later disputed. Nevertheless, people were able to develop domestic breeds of animals as well as crops through artificial selection. The inheritance of acquired traits also formed a part of early Lamarckian ideas on evolution.

During the 18th century, Dutch microscopist Antonie van Leeuwenhoek (1632–1723) discovered "animalcules" in the sperm of humans and other animals. Some scientists speculated they saw a "little man" (homunculus) inside each sperm. These scientists formed a school of thought known as the "spermists". They contended the only contributions of the female to the next generation were the womb in which the homunculus grew, and prenatal influences of the womb. An opposing school of thought, the ovists, believed that the future human was in the egg, and that sperm merely stimulated the growth of the egg. Ovists thought women carried eggs containing boy and girl children, and that the gender of the offspring was determined well before conception.

Gregor Mendel: Father of Genetics

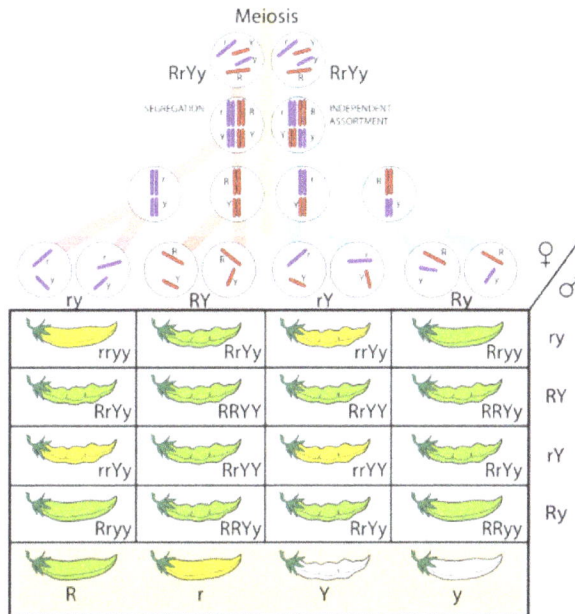

Table showing how the genes exchange according to segregation or independent assortment during meiosis and how this translates into Mendel's laws

The idea of particulate inheritance of genes can be attributed to the Moravian monk Gregor Mendel who published his work on pea plants in 1865. However, his work was not widely known and was rediscovered in 1901. It was initially assumed the Mendelian inheritance only accounted for large (qualitative) differences, such as those seen by Mendel in his pea plants—and the idea of additive effect of (quantitative) genes was not realised until R. A. Fisher's (1918) paper, "The Correlation Between Relatives on the Supposition of Mendelian Inheritance" Mendel's overall contribution gave scientists a useful overview that traits were inheritable. His pea plant demonstration became the foundation of the study of Mendelian Traits. These traits can be traced on a single locus.

Modern Development of Genetics and Heredity

In the 1930s, work by Fisher and others resulted in a combination of Mendelian and biometric schools into the modern evolutionary synthesis. The modern synthesis bridged the gap between experimental geneticists and naturalists; and between both and palaeontologists, stating that:

1. All evolutionary phenomena can be explained in a way consistent with known genetic mechanisms and the observational evidence of naturalists.

2. Evolution is gradual: small genetic changes, recombination ordered by natural selection. Discontinuities amongst species (or other taxa) are explained as originating gradually through geographical separation and extinction (not saltation).

3. Selection is overwhelmingly the main mechanism of change; even slight advantages are important when continued. The object of selection is the phenotype in its surrounding environment. The role of genetic drift is equivocal; though strongly supported initially

by Dobzhansky, it was downgraded later as results from ecological genetics were obtained.

4. The primacy of population thinking: the genetic diversity carried in natural populations is a key factor in evolution. The strength of natural selection in the wild was greater than expected; the effect of ecological factors such as niche occupation and the significance of barriers to gene flow are all important.

The idea that speciation occurs after populations are reproductively isolated has been much debated. In plants, polyploidy must be included in any view of speciation. Formulations such as 'evolution consists primarily of changes in the frequencies of alleles between one generation and another' were proposed rather later. The traditional view is that developmental biology ('evo-devo') played little part in the synthesis, but an account of Gavin de Beer's work by Stephen Jay Gould suggests he may be an exception.

Almost all aspects of the synthesis have been challenged at times, with varying degrees of success. There is no doubt, however, that the synthesis was a great landmark in evolutionary biology. It cleared up many confusions, and was directly responsible for stimulating a great deal of research in the post-World War II era.

Trofim Lysenko however caused a backlash of what is now called Lysenkoism in the Soviet Union when he emphasised Lamarckian ideas on the inheritance of acquired traits. This movement affected agricultural research and led to food shortages in the 1960s and seriously affected the USSR.

Common Genetic Disorders

- Down syndrome

- sickle cell disease

- Phenylketonuria (PKU)

- Haemophilia

Types

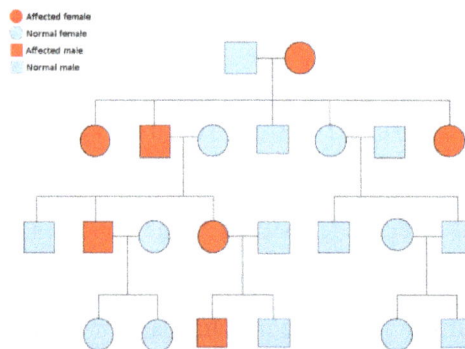

An example pedigree chart of an autosomal dominant disorder.

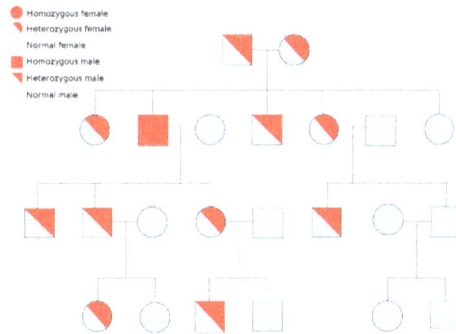

An example pedigree chart of an autosomal recessive disorder.

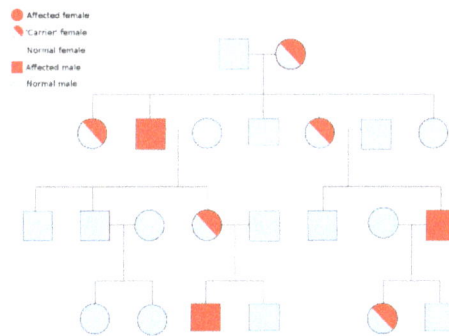

An example pedigree chart of a sex-linked disorder (the gene is on the X chromosome)

Dominant and Recessive Alleles

An allele is said to be dominant if it is always expressed in the appearance of an organism (phenotype) provided that at least one copy of it is present. For example, in peas the allele for green pods, *G*, is dominant to that for yellow pods, *g*. Thus pea plants with the pair of alleles *either GG* (homozygote) *or Gg* (heterozygote) will have green pods. The allele for yellow pods is recessive. The effects of this allele are only seen when it is present in both chromosomes, *gg* (homozygote).

The description of a mode of biological inheritance consists of three main categories:

1. Number of Involved Loci

- Monogenetic (also called "simple") – one locus

- Oligogenetic – few loci

- Polygenetic – many loci

2. Involved Chromosomes

- Autosomal – loci are not situated on a sex chromosome

- Gonosomal – loci are situated on a sex chromosome

        o   X-chromosomal – loci are situated on the X-chromosome (the more common case)

        o   Y-chromosomal – loci are situated on the Y-chromosome

- Mitochondrial – loci are situated on the mitochondrial DNA

3. Correlation Genotype–phenotype

- Dominant

- Intermediate (also called "codominant")

- Recessive

- Overdominant

- Underdominant

These three categories are part of every exact description of a mode of inheritance in the above order. In addition, more specifications may be added as follows:

4. Coincidental and Environmental Interactions

- Penetrance

        o   Complete

        o   Incomplete (percentual number)

- Expressivity

        o   Invariable

        o   Variable

- Heritability (in polygenetic and sometimes also in oligogenetic modes of inheritance)

- Maternal or paternal imprinting phenomena

5. Sex-linked Interactions

- Sex-linked inheritance (gonosomal loci)

- Sex-limited phenotype expression (e.g., cryptorchism)

- Inheritance through the maternal line (in case of mitochondrial DNA loci)

- Inheritance through the paternal line (in case of Y-chromosomal loci)

6. Locus–locus interactions

- Epistasis with other loci (e.g., overdominance)

- Gene coupling with other loci

- Homozygotous lethal factors

- Semi-lethal factors

Determination and description of a mode of inheritance is achieved primarily through statistical analysis of pedigree data. In case the involved loci are known, methods of molecular genetics can also be employed.

Chromosome

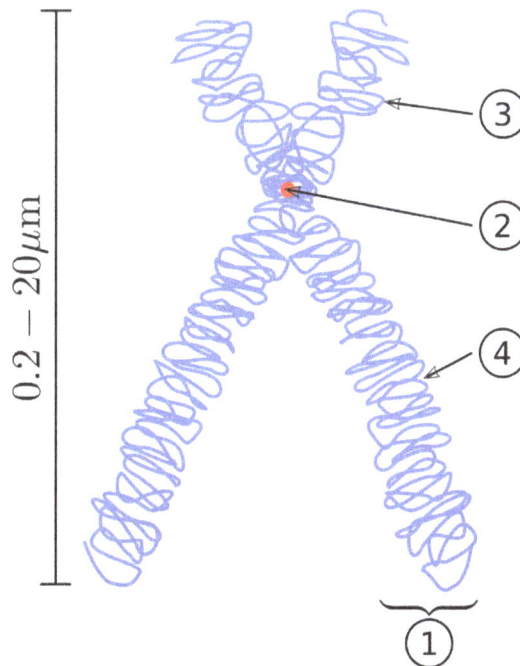

Diagram of a replicated and condensed metaphase eukaryotic chromosome. (1) Chromatid – one of the two identical parts of the chromosome after S phase. (2) Centromere – the point where the two chromatids touch. (3) Short arm. (4) Long arm.

A chromosome is a packaged and organized structure containing most of the DNA of a living organism. DNA is not usually found on its own, but rather is structured in long strands which are wrapped around protein complexes called nucleosomes, which consist of proteins called histones. The DNA in chromosomes serves as the source for transcription. Most eukaryotic cells have a set of chromosomes (46 in humans) with the genetic material spread among them.

During most of the duration of the cell cycle, a chromosome consists of one long (its width to length ratio is about 1:65,000,000) double-helix DNA molecule (with associated proteins). During

S phase, the chromosome gets replicated, resulting in an X-shaped structure called a metaphase chromosome. Both the original and the newly copied DNA are now called chromatids. The two "sister" chromatids are joined together at a protein junction called a centromere (forming the X-shaped structure).

Chromosomes are normally visible under a light microscope only when the cell is undergoing mitosis (cell division). Even then, the full chromosome containing both joined sister chromatids becomes visible only during a sequence of mitosis known as metaphase (when chromosomes align together, attached to the mitotic spindle and prepare to divide). This DNA and its associated proteins and macromolecules is collectively known as chromatin, which is further packaged along with its associated molecules into a discrete structure called a nucleosome. Chromatin is present in most cells, with a few exceptions, for example, red blood cells. Occurring only in the nucleus of eukaryotic cells, chromatin composes the vast majority of all DNA, except for a small amount inherited maternally, which is found in mitochondria.

In prokaryotic cells, chromatin occurs free-floating in cytoplasm, as these cells lack organelles and a defined nucleus. Bacteria also lack histones. The main information-carrying macromolecule is a single piece of coiled double-helix DNA, containing many genes, regulatory elements and other noncoding DNA. The DNA-bound macromolecules are proteins, which serve to package the DNA and control its functions. Chromosomes vary widely between different organisms. Some species such as certain bacteria also contain plasmids or other extrachromosomal DNA. These are circular structures in the cytoplasm, which contain cellular DNA and play a role in horizontal gene transfer.

Compaction of the duplicated chromosomes during cell division (mitosis or meiosis) results either in a four-arm structure (pictured above) if the centromere is located in the middle of the chromosome or a two-arm structure if the centromere is located near one of the ends. Chromosomal recombination during meiosis and subsequent sexual reproduction plays a significant role in genetic diversity. If these structures are manipulated incorrectly, through processes known as chromosomal instability and translocation, the cell may undergo mitotic catastrophe and die, or it may unexpectedly evade apoptosis leading to the progression of cancer.

In prokaryotes and viruses, the DNA is often densely packed and organized: in the case of archaea, by homologs to eukaryotic histones, and in the case of bacteria, by histone-like proteins. Small circular genomes called plasmids are often found in bacteria and also in mitochondria and chloroplasts, reflecting their bacterial origins.

Some authors, as in this article, use the term chromosome in a wider sense, to refer to the individualized portions of chromatin in cells, either visible or not under light microscopy. However, others use the concept in a narrower sense, to refer to the individualized portions of chromatin during cell division, visible under light microscopy due to high condensation.

History of Discovery

The word *chromosome* comes from the Greek *chroma*, "co-lour" and *soma*, "body", describing their strong staining by particular dyes.

Schleiden, Virchow and Bütschli were among the first scientists who recognized the structures now so familiar to everyone as chromosomes. The term was coined by von Waldeyer-Hartz, referring to the term chromatin, which was introduced by Walther Flemming.

Walter Sutton (left) and Theodor Boveri (right) independently developed
the chromosome theory of inheritance in 1902.

In a series of experiments beginning in the mid-1880s, Theodor Boveri gave the definitive demonstration that chromosomes are the vectors of heredity. His two principles were the *continuity* of chromosomes and the *individuality* of chromosomes.It is the second of these principles that was so original. Wilhelm Roux suggested that each chromosome carries a different genetic load. Boveri was able to test and confirm this hypothesis. Aided by the rediscovery at the start of the 1900s of Gregor Mendel's earlier work, Boveri was able to point out the connection between the rules of inheritance and the behaviour of the chromosomes. Boveri influenced two generations of American cytologists: Edmund Beecher Wilson, Nettie Stevens, Walter Sutton and Theophilus Painter were all influenced by Boveri (Wilson, Stevens, and Painter actually worked with him).

In his famous textbook *The Cell in Development and Heredity*, Wilson linked together the independent work of Boveri and Sutton (both around 1902) by naming the chromosome theory of inheritance the Boveri–Sutton chromosome theory (the names are sometimes reversed). Ernst Mayr remarks that the theory was hotly contested by some famous geneticists: William Bateson, Wilhelm Johannsen, Richard Goldschmidt and T.H. Morgan, all of a rather dogmatic turn of mind. Eventually, complete proof came from chromosome maps in Morgan's own lab.

The number of human chromosomes was published in 1923 by Theophilus Painter. By inspection through the microscope he counted 24 pairs which would mean 48 chromosomes. His error was copied by others and it was not until 1956 that the true number, 46, was determined by Indonesia-born cytogeneticist Joe Hin Tjio.

Prokaryotes

The prokaryotes – bacteria and archaea – typically have a single circular chromosome, but many variations exist. The chromosomes of most bacteria, which some authors prefer to call genophores, can range in size from only 130,000 base pairs in the endosymbiotic bacteria *Candidatus Hodgkinia cicadicola* and *Candidatus Tremblaya princeps*, to over 14,000,000 base pairs in the soil-dwelling bacterium *Sorangium cellulosum*. Spirochaetes of the genus *Borrelia* are a notable exception to this arrangement, with bacteria such as *Borrelia burgdorferi*, the cause of Lyme disease, containing a single *linear* chromosome.

Structure in Sequences

Prokaryotic chromosomes have less sequence-based structure than eukaryotes. Bacteria typically have a one-point (the origin of replication) from which replication starts, whereas some archaea contain multiple replication origins. The genes in prokaryotes are often organized in operons, and do not usually contain introns, unlike eukaryotes.

DNA Packaging

Prokaryotes do not possess nuclei. Instead, their DNA is organized into a structure called the nucleoid. The nucleoid is a distinct structure and occupies a defined region of the bacterial cell. This structure is, however, dynamic and is maintained and remodeled by the actions of a range of histone-like proteins, which associate with the bacterial chromosome. In archaea, the DNA in chromosomes is even more organized, with the DNA packaged within structures similar to eukaryotic nucleosomes.

Bacterial chromosomes tend to be tethered to the plasma membrane of the bacteria. In molecular biology application, this allows for its isolation from plasmid DNA by centrifugation of lysed bacteria and pelleting of the membranes (and the attached DNA).

Prokaryotic chromosomes and plasmids are, like eukaryotic DNA, generally supercoiled. The DNA must first be released into its relaxed state for access for transcription, regulation, and replication.

Eukaryotes

In eukaryotes, nuclear chromosomes are packaged by proteins into a condensed structure called chromatin. This allows the very long DNA molecules to fit into the cell nucleus. The structure of chromosomes and chromatin varies through the cell cycle. Chromosomes are even more condensed than chromatin and are an essential unit for cellular division. Chromosomes must be replicated, divided, and passed successfully to their daughter cells so as to ensure the genetic diversity and survival of their progeny. Chromosomes may exist as either duplicated or unduplicated. Unduplicated chromosomes are single double helixes, whereas duplicated chromosomes contain two identical copies (called chromatids or sister chromatids) joined by a centromere.

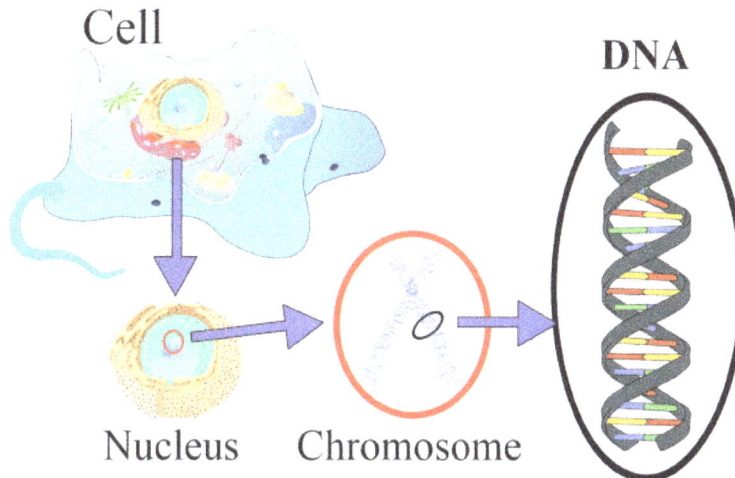

Organization of DNA in a eukaryotic cell.

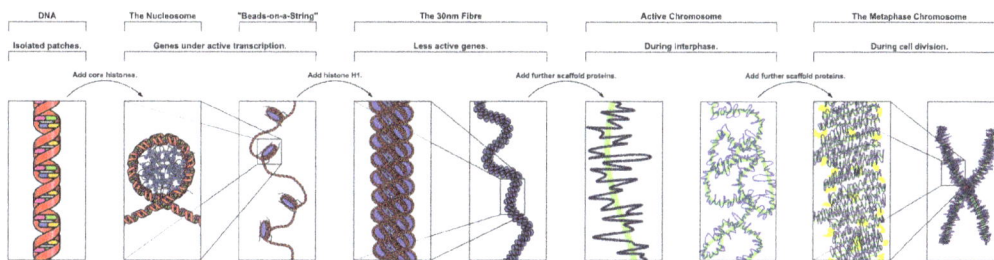

The major structures in DNA compaction: DNA, the nucleosome, the 10 nm "beads-on-a-string" fibre, the 30 nm fibre and the metaphase chromosome.

Eukaryotes (cells with nuclei such as those found in plants, fungi, and animals) possess multiple large linear chromosomes contained in the cell's nucleus. Each chromosome has one centromere, with one or two arms projecting from the centromere, although, under most circumstances, these arms are not visible as such. In addition, most eukaryotes have a small circular mitochondrial genome, and some eukaryotes may have additional small circular or linear cytoplasmic chromosomes.

In the nuclear chromosomes of eukaryotes, the uncondensed DNA exists in a semi-ordered structure, where it is wrapped around histones (structural proteins), forming a composite material called chromatin.

Chromatin

Chromatin is the complex of DNA and protein found in the eukaryotic nucleus, which packages chromosomes. The structure of chromatin varies significantly between different stages of the cell cycle, according to the requirements of the DNA.

Interphase Chromatin

During interphase (the period of the cell cycle where the cell is not dividing), two types of chromatin can be distinguished:

- Euchromatin, which consists of DNA that is active, e.g., being expressed as protein.

- Heterochromatin, which consists of mostly inactive DNA. It seems to serve structural purposes during the chromosomal stages. Heterochromatin can be further distinguished into two types:

  o *Constitutive heterochromatin*, which is never expressed. It is located around the centromere and usually contains repetitive sequences.

  o *Facultative heterochromatin*, which is sometimes expressed.

Metaphase Chromatin and Division

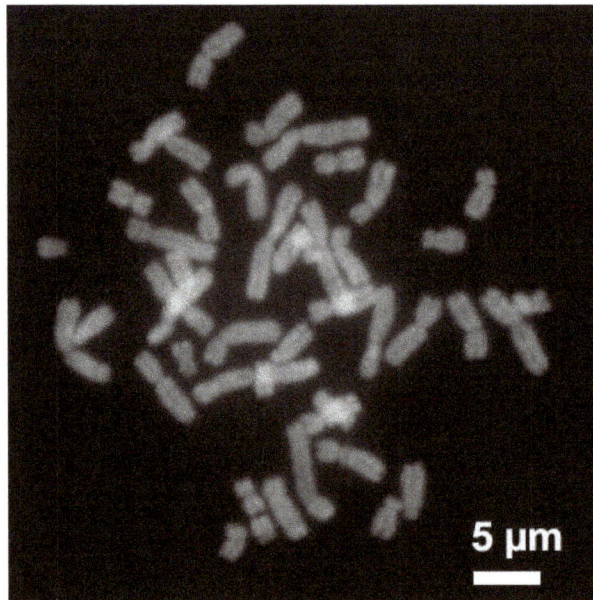

Human chromosomes during metaphase

In the early stages of mitosis or meiosis (cell division), the chromatin double helix become more and more condensed. They cease to function as accessible genetic material (transcription stops) and become a compact transportable form. This compact form makes the individual chromosomes visible, and they form the classic four arm structure, a pair of sister chromatids attached to each other at the centromere. The shorter arms are called *p arms* (from the French *petit*, small) and the longer arms are called *q arms* (q follows p in the Latin alphabet; q-g "grande"; alternatively it is sometimes said q is short for *queue* meaning tail in French). This is the only natural context in which individual chromosomes are visible with an optical microscope.

Mitotic metaphase chromosomes are best described by a linearly organized longitudinally compressed array of consecutive chromatin loops.

During mitosis, microtubules grow from centrosomes located at opposite ends of the cell and also attach to the centromere at specialized structures called kinetochores, one of which is present on each sister chromatid. A special DNA base sequence in the region of the kineto-chores provides, along with special proteins, longer-lasting attachment in this region. The microtubules then pull the chromatids apart toward the centrosomes, so that each daughter cell inherits one set of chromatids. Once the cells have divided, the chromatids are uncoiled and DNA can again be transcribed. In spite of their appearance, chromosomes are structural-ly highly condensed, which enables these giant DNA structures to be contained within a cell nucleus.

Human Chromosomes

Chromosomes in humans can be divided into two types: autosomes (body chromosome(s)) and allosome (sex chromosome(s)). Certain genetic traits are linked to a person's sex and are passed on through the sex chromosomes. The autosomes contain the rest of the genetic hered-itary information. All act in the same way during cell division. Human cells have 23 pairs of chromosomes (22 pairs of autosomes and one pair of sex chromosomes), giving a total of 46 per cell. In addition to these, human cells have many hundreds of copies of the mitochondrial genome. Sequencing of the human genome has provided a great deal of information about each of the chromosomes. Below is a table compiling statistics for the chromosomes, based on the Sanger Institute's human genome information in the Vertebrate Genome Annotation (VEGA) database. Number of genes is an estimate, as it is in part based on gene predictions. Total chromosome length is an estimate as well, based on the estimated size of unsequenced heterochromatin regions.

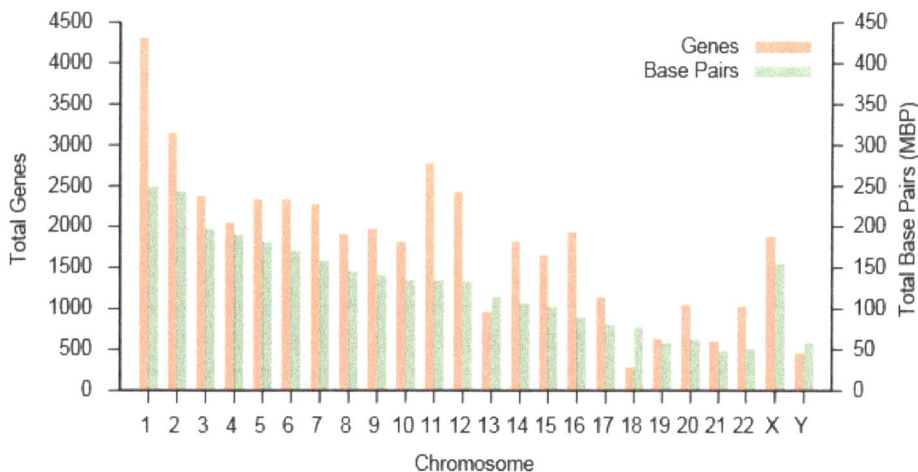

Estimated number of genes and base pairs (in mega base pairs) on each human chromosome

Chromosome	Genes	Total base pairs	% of bases	Sequenced base pairs
1	2000	247,199,719	8.0	224,999,719
2	1300	242,751,149	7.9	237,712,649
3	1000	199,446,827	6.5	194,704,827

4	1000	191,263,063	6.2	187,297,063
5	900	180,837,866	5.9	177,702,766
6	1000	170,896,993	5.5	167,273,993
7	900	158,821,424	5.2	154,952,424
8	700	146,274,826	4.7	142,612,826
9	800	140,442,298	4.6	120,312,298
10	700	135,374,737	4.4	131,624,737
11	1300	134,452,384	4.4	131,130,853
12	1100	132,289,534	4.3	130,303,534
13	300	114,127,980	3.7	95,559,980
14	800	106,360,585	3.5	88,290,585
15	600	100,338,915	3.3	81,341,915
16	800	88,822,254	2.9	78,884,754
17	1200	78,654,742	2.6	77,800,220
18	200	76,117,153	2.5	74,656,155
19	1500	63,806,651	2.1	55,785,651
20	500	62,435,965	2.0	59,505,254
21	200	46,944,323	1.5	34,171,998
22	500	49,528,953	1.6	34,893,953
X (sex chromosome)	800	154,913,754	5.0	151,058,754
Y (sex chromosome)	50	57,741,652	1.9	25,121,652
Total	21,000	3,079,843,747	100.0	2,857,698,560

Number in Various Organisms

In Eukaryotes

These tables give the total number of chromosomes (including sex chromosomes) in a cell nucleus. For example, human cells are diploid and have 22 different types of autosome, each present as two copies, and two sex chromosomes. This gives 46 chromosomes in total. Other organisms have more than two copies of their chromosome types, such as bread wheat, which is *hexaploid* and has six copies of seven different chromosome types – 42 chromosomes in total.

Chromosome numbers in some plants	
Plant Species	#
Arabidopsis thaliana (diploid)	10
Rye (diploid)	14
Maize (diploid or palaeotetraploid)	20
Einkorn wheat (diploid)	14
Durum wheat (tetraploid)	28
Bread wheat (hexaploid)	42
Cultivated tobacco (tetraploid)	48
Adder's tongue fern (diploid)	approx. 1,200

Chromosome numbers (2n) in some animals			
Species	#	Species	#
Common fruit fly	8	Guinea pig	64
Guppy (*poecilia reticulata*)	46	Garden snail	54
Earthworm (*Octodrilus complanatus*)	36		
Pill millipede (*Arthrosphaera fumosa*)	30		
Tibetan fox	36		
Domestic cat	38	Domestic pig	38
Laboratory mouse	40	Laboratory rat	42
Rabbit (*Oryctolagus cuniculus*)	44	Syrian hamster	44
Hares	48	Human	46
Gorillas, chimpanzees	48	Domestic sheep	54
Elephants	56	Cow	60
Donkey	62	Horse	64
Dog	78		
Hedgehog	90		
Kingfisher	132		
Goldfish	100-104	Silkworm	56

Chromosome numbers in other organisms			
Species	Large Chromosomes	Intermediate Chromosomes	Microchromosomes
Trypanosoma brucei	11	6	~100
Domestic pigeon (*Columba livia domestics*)	18	-	59-63
Chicken	8	2 sex chromosomes	60

Normal members of a particular eukaryotic species all have the same number of nuclear chromosomes. Other eukaryotic chromosomes, i.e., mitochondrial and plasmid-like small chromosomes, are much more variable in number, and there may be thousands of copies per cell.

Asexually reproducing species have one set of chromosomes, which are the same in all body cells. However, asexual species can be either haploid or diploid.

Sexually reproducing species have somatic cells (body cells), which are diploid [2n] having two sets of chromosomes (23 pairs in humans with one set of 23 chromosomes from each parent), one set from the mother and one from the father. Gametes, reproductive cells, are haploid [n]: They have one set of chromosomes. Gametes are produced by meiosis of a diploid germ line cell. During meiosis, the matching chromosomes of father and mother can exchange small parts of themselves (crossover), and thus create new chromosomes that are not inherited solely from either parent. When a male and a female gamete merge (fertilization), a new diploid organism is formed.

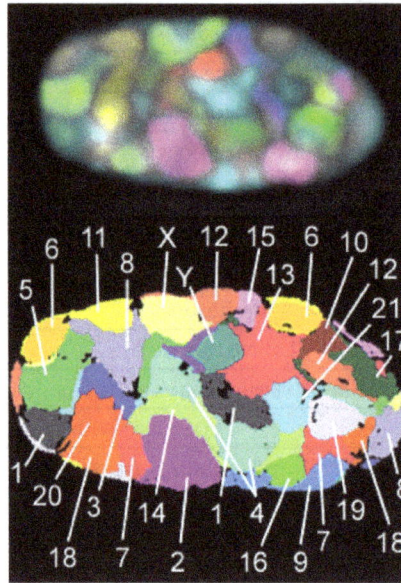

The 23 human chromosome territories during prometaphase in fibroblast cells.

Some animal and plant species are polyploid [Xn]: They have more than two sets of homologous chromosomes. Plants important in agriculture such as tobacco or wheat are often polyploid, compared to their ancestral species. Wheat has a haploid number of seven chromosomes, still seen in some cultivars as well as the wild progenitors. The more-common pasta and bread wheats are polyploid, having 28 (tetraploid) and 42 (hexaploid) chromosomes, compared to the 14 (diploid) chromosomes in the wild wheat.

In Prokaryotes

Prokaryote species generally have one copy of each major chromosome, but most cells can easily survive with multiple copies. For example, *Buchnera*, a symbiont of aphids has multiple copies of its chromosome, ranging from 10–400 copies per cell. However, in some large bacteria, such as *Epulopiscium fishelsoni* up to 100,000 copies of the chromosome can be present. Plasmids and plasmid-like small chromosomes are, as in eukaryotes, highly variable in copy number. The number of plasmids in the cell is almost entirely determined by the rate of division of the plasmid – fast division causes high copy number.

Karyotype

In general, the karyotype is the characteristic chromosome complement of a eukaryote species. The preparation and study of karyotypes is part of cytogenetics.

Although the replication and transcription of DNA is highly standardized in eukaryotes, *the same cannot be said for their karyotypes*, which are often highly variable. There may be variation between species in chromosome number and in detailed organization. In some cases, there is significant variation within species. Often there is:

    1. variation between the two sexes

    2. variation between the germ-line and soma (between gametes and the rest of the body)

3. variation between members of a population, due to balanced genetic polymorphism

4. geographical variation between races

5. mosaics or otherwise abnormal individuals.

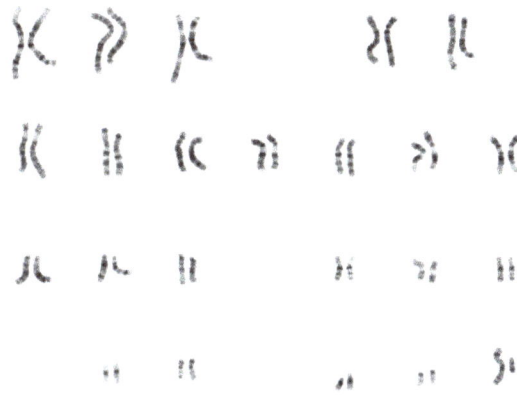

Karyogram of a human male

Also, variation in karyotype may occur during development from the fertilized egg.

The technique of determining the karyotype is usually called *karyotyping*. Cells can be locked part-way through division (in metaphase) in vitro (in a reaction vial) with colchicine. These cells are then stained, photographed, and arranged into a *karyogram*, with the set of chromosomes arranged, autosomes in order of length, and sex chromosomes (here X/Y) at the end.

Like many sexually reproducing species, humans have special gonosomes (sex chromosomes, in contrast to autosomes). These are XX in females and XY in males.

Historical Note

Investigation into the human karyotype took many years to settle the most basic question: *How many chromosomes does a normal diploid human cell contain?* In 1912, Hans von Winiwarter reported 47 chromosomes in spermatogonia and 48 in oogonia, concluding an XX/XO sex determination mechanism. Painter in 1922 was not certain whether the diploid number of man is 46 or 48, at first favouring 46. He revised his opinion later from 46 to 48, and he correctly insisted on humans having an XX/XY system.

New techniques were needed to definitively solve the problem:

1. Using cells in culture

2. Arresting mitosis in metaphase by a solution of colchicine

3. Pretreating cells in a hypotonic solution 0.075 M KCl, which swells them and spreads the chromosomes

4. Squashing the preparation on the slide forcing the chromosomes into a single plane

5. Cutting up a photomicrograph and arranging the result into an indisputable karyogram.

It took until 1954 before the human diploid number was confirmed as 46. Considering the techniques of Winiwarter and Painter, their results were quite remarkable. Chimpanzees (the closest living relatives to modern humans) have 48 chromosomes (as well as the other great apes: in humans two chromosomes fused to form chromosome 2).

Aberrations

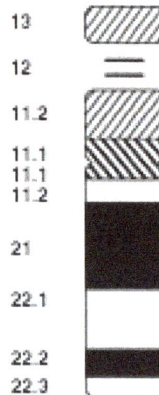

In Down syndrome, there are three copies of chromosome 21

Chromosomal aberrations are disruptions in the normal chromosomal content of a cell and are a major cause of genetic conditions in humans, such as Down syndrome, although most aberrations have little to no effect. Some chromosome abnormalities do not cause disease in carriers, such as translocations, or chromosomal inversions, although they may lead to a higher chance of bearing a child with a chromosome disorder. Abnormal numbers of chromosomes or chromosome sets, called aneuploidy, may be lethal or may give rise to genetic disorders. Genetic counseling is offered for families that may carry a chromosome rearrangement.

The gain or loss of DNA from chromosomes can lead to a variety of genetic disorders. Human examples include:

- Cri du chat, which is caused by the deletion of part of the short arm of chromosome 5. "Cri du chat" means "cry of the cat" in French; the condition was so-named because affected babies make high-pitched cries that sound like those of a cat. Affected individuals have wide-set eyes, a small head and jaw, moderate to severe mental health problems, and are very short.

- Down's syndrome, the most common trisomy, usually caused by an extra copy of chromosome 21 (trisomy 21). Characteristics include decreased muscle tone, stockier build, asymmetrical skull, slanting eyes and mild to moderate developmental disability.

- Edwards syndrome, or trisomy-18, the second most common trisomy. Symptoms include motor retardation, developmental disability and numerous congenital anomalies causing serious health problems. Ninety percent of those affected die in infancy. They have characteristic clenched hands and overlapping fingers.

- Isodicentric 15, also called idic(15), partial tetrasomy 15q, or inverted duplication 15 (inv dup 15).

- Jacobsen syndrome, which is very rare. It is also called the terminal 11q deletion disorder. Those affected have normal intelligence or mild developmental disability, with poor expressive language skills. Most have a bleeding disorder called Paris-Trousseau syndrome.

- Klinefelter syndrome (XXY). Men with Klinefelter syndrome are usually sterile and tend to be taller and have longer arms and legs than their peers. Boys with the syndrome are often shy and quiet and have a higher incidence of speech delay and dyslexia. Without testosterone treatment, some may develop gynecomastia during puberty.

- Patau Syndrome, also called D-Syndrome or trisomy-13. Symptoms are somewhat similar to those of trisomy-18, without the characteristic folded hand.

- Small supernumerary marker chromosome. This means there is an extra, abnormal chromosome. Features depend on the origin of the extra genetic material. Cat-eye syndrome and isodicentric chromosome 15 syndrome (or Idic15) are both caused by a supernumerary marker chromosome, as is Pallister–Killian syndrome.

- Triple-X syndrome (XXX). XXX girls tend to be tall and thin and have a higher incidence of dyslexia.

- Turner syndrome (X instead of XX or XY). In Turner syndrome, female sexual characteristics are present but underdeveloped. Females with Turner syndrome often have a short stature, low hairline, abnormal eye features and bone development and a "caved-in" appearance to the chest.

- Wolf–Hirschhorn syndrome, which is caused by partial deletion of the short arm of chromosome 4. It is characterized by growth retardation, delayed motor skills development, "Greek Helmet" facial features, and mild to profound mental health problems.

- XYY syndrome. XYY boys are usually taller than their siblings. Like XXY boys and XXX girls, they are more likely to have learning difficulties.

Sperm Aneuploidy

Exposure of males to certain lifestyle, environmental and/or occupational hazards may increase the risk of aneuploid spermatozoa. In particular, risk of aneuploidy is increased by tobacco smoking, and occupational exposure to benzene, insecticides, and perfluorinated compounds. Increased aneuploidy is often associated with increased DNA damage in spermatozoa.

RNA

Ribonucleic acid (RNA) is a polymeric molecule essential in various biological roles in coding, decoding, regulation, and expression of genes. RNA and DNA are nucleic acids, and, along with proteins and carbohydrates, constitute the four major macromolecules essential for all known forms of life. Like DNA, RNA is assembled as a chain of nucleotides, but unlike DNA it is more often found in nature as a single-strand folded onto itself, rather than a paired double-strand. Cellular organisms use messenger RNA (*mRNA*) to convey genetic information (using the letters G, U, A, and C to denote the nitrogenous bases guanine, uracil, adenine, and cytosine) that directs synthesis of specific proteins. Many viruses encode their genetic information using an RNA genome.

A hairpin loop from a pre-mRNA. Highlighted are the nucleobases (green) and the ribose-phosphate backbone (blue). Note that this is a single strand of RNA that folds back upon itself.

Some RNA molecules play an active role within cells by catalyzing biological reactions, controlling gene expression, or sensing and communicating responses to cellular signals. One of these active processes is protein synthesis, a universal function wherein mRNA molecules direct the assembly of proteins on ribosomes. This process uses transfer RNA (*tRNA*) molecules to deliver amino acids to the ribosome, where ribosomal RNA (*rRNA*) then links amino acids together to form proteins.

Comparison with DNA

Three-dimensional representation of the 50S ribosomal subunit. Ribosomal RNA is in ochre, proteins in blue. The active site is a small segment of rRNA, indicated in red.

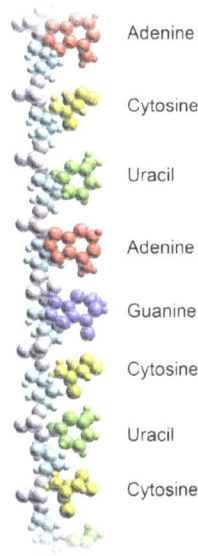

RNA Molecule

Bases in an RNA molecule.

The chemical structure of RNA is very similar to that of DNA, but differs in three main ways:

- Unlike double-stranded DNA, RNA is a single-stranded molecule in many of its biological roles and has a much shorter chain of nucleotides. However, RNA can, by complementary base pairing, form intrastrand (i.e., single-strand) double helixes, as in tRNA.

- While DNA contains *deoxyribose*, RNA contains *ribose* (in deoxyribose there is no hydroxyl group attached to the pentose ring in the 2' position). These hydroxyl groups make RNA less stable than DNA because it is more prone to hydrolysis.

- The complementary base to adenine in DNA is thymine, whereas in RNA, it is uracil, which is an unmethylated form of thymine.

Like DNA, most biologically active RNAs, including mRNA, tRNA, rRNA, snRNAs, and other non-coding RNAs, contain self-complementary sequences that allow parts of the RNA to fold and pair with itself to form double helices. Analysis of these RNAs has revealed that they are highly structured. Unlike DNA, their structures do not consist of long double helices, but rather collections of short helices packed together into structures akin to proteins. In this fashion, RNAs can achieve chemical catalysis (like enzymes). For instance, determination of the structure of the ribosome—an enzyme that catalyzes peptide bond formation—revealed that its active site is composed entirely of RNA.

Structure

Each nucleotide in RNA contains a ribose sugar, with carbons numbered 1' through 5'. A base is attached to the 1' position, in general, adenine (A), cytosine (C), guanine (G), or uracil (U). Adenine and guanine are purines, cytosine and uracil are pyrimidines. A phosphate group is attached to the 3' position of one ribose and the 5' position of the next. The phosphate groups have a negative charge each, making RNA a charged molecule (polyanion). The bases form hydrogen bonds between cytosine and guanine, between adenine and uracil and between guanine and uracil. However, other interactions are possible, such as a group of adenine bases binding to each other in a bulge, or the GNRA tetraloop that has a guanine–adenine base-pair.

Watson-Crick base pairs in a siRNA (hydrogen atoms are not shown)

An important structural feature of RNA that distinguishes it from DNA is the presence of a hydroxyl group at the 2' position of the ribose sugar. The presence of this functional group causes the helix to mostly adopt the A-form geometry, although in single strand dinucleotide contexts, RNA can rarely also adopt the B-form most commonly observed in DNA. The A-form geometry results in a very deep and narrow major groove and a shallow and wide minor groove. A second consequence of the presence of the 2'-hydroxyl group is that in conformationally flexible regions of an RNA molecule (that is, not involved in formation of a double helix), it can chemically attack the adjacent phosphodiester bond to cleave the backbone.

T. thermophila telomerase RNA

Secondary structure of a telomerase RNA.

RNA is transcribed with only four bases (adenine, cytosine, guanine and uracil), but these bases and attached sugars can be modified in numerous ways as the RNAs mature. Pseudouridine (Ψ), in which the linkage between uracil and ribose is changed from a C–N bond to a C–C bond, and ribothymidine (T) are found in various places (the most notable ones being in the TΨC loop of tRNA). Another notable modified base is hypoxanthine, a deaminated adenine base whose nucleoside is called inosine (I). Inosine plays a key role in the wobble hypothesis of the genetic code.

There are more than 100 other naturally occurring modified nucleosides, The greatest structural diversity of modifications can be found in tRNA, while pseudouridine and nucleosides with 2'-O-methylribose often present in rRNA are the most common. The specific roles of many of these modifications in RNA are not fully understood. However, it is notable that, in ribosomal RNA, many of the post-transcriptional modifications occur in highly functional regions, such as the peptidyl transferase center and the subunit interface, implying that they are important for normal function.

The functional form of single-stranded RNA molecules, just like proteins, frequently requires a specific tertiary structure. The scaffold for this structure is provided by secondary structural elements that are hydrogen bonds within the molecule. This leads to several recognizable "domains" of secondary structure like hairpin loops, bulges, and internal loops. Since RNA is charged, metal ions such as Mg^{2+} are needed to stabilise many secondary and tertiary structures.

The naturally occurring enantiomer of RNA is D-RNA composed of D-ribonucleotides. All chirality centers are located in the D-ribose. By the use of L-ribose or rather L-ribonucleotides, L-RNA can be synthesized. L-RNA is much more stable against degradation by RNase.

Like other structured biopolymers such as proteins, one can define topology of a folded RNA molecule. This is often done based on arrangement of intra-chain contacts within a folded RNA, termed as circuit topology.

Synthesis

Synthesis of RNA is usually catalyzed by an enzyme—RNA polymerase—using DNA as a template, a process known as transcription. Initiation of transcription begins with the binding of the enzyme to a promoter sequence in the DNA (usually found "upstream" of a gene). The DNA double helix is unwound by the helicase activity of the enzyme. The enzyme then progresses along the template strand in the 3' to 5' direction, synthesizing a complementary RNA molecule with elongation occurring in the 5' to 3' direction. The DNA sequence also dictates where termination of RNA synthesis will occur.

Primary transcript RNAs are often modified by enzymes after transcription. For example, a poly(A) tail and a 5' cap are added to eukaryotic pre-mRNA and introns are removed by the spliceosome.

There are also a number of RNA-dependent RNA polymerases that use RNA as their template for synthesis of a new strand of RNA. For instance, a number of RNA viruses (such as poliovirus) use this type of enzyme to replicate their genetic material. Also, RNA-dependent RNA polymerase is part of the RNA interference pathway in many organisms.

Types of RNA

Overview

Messenger RNA (mRNA) is the RNA that carries information from DNA to the ribosome, the sites of protein synthesis (translation) in the cell. The coding sequence of the mRNA determines the amino acid sequence in the protein that is produced. However, many RNAs do not code for protein (about 97% of the transcriptional output is non-protein-coding in eukaryotes).

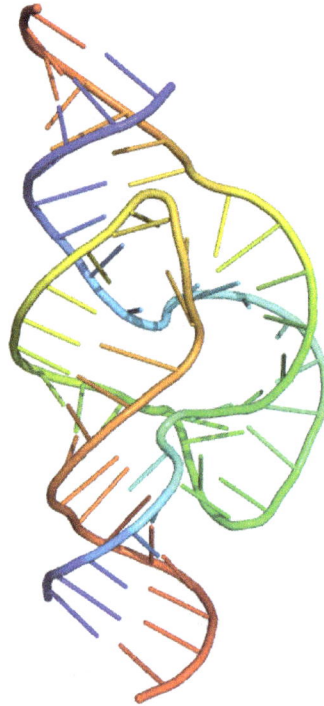

Structure of a hammerhead ribozyme, a ribozyme that cuts RNA

These so-called non-coding RNAs ("ncRNA") can be encoded by their own genes (RNA genes), but can also derive from mRNA introns. The most prominent examples of non-coding RNAs are transfer RNA (tRNA) and ribosomal RNA (rRNA), both of which are involved in the process of translation. There are also non-coding RNAs involved in gene regulation, RNA processing and other roles. Certain RNAs are able to catalyse chemical reactions such as cutting and ligating other RNA molecules, and the catalysis of peptide bond formation in the ribosome; these are known as ribozymes.

In Length

According to the length of RNA chain, RNA includes small RNA and long RNA. Usually, small RNAs are <200 nt in length, and long RNA are >200 nt. Long RNAs, also called large RNAs, mainly include long non-coding RNA (lncRNA) and mRNA. Small RNAs mainly include 5.8S ribosomal RNA (rRNA), 5S rRNA, transfer RNA (tRNA), microRNA (miRNA), small interfering RNA (siRNA), small nucleolar RNA (snoRNAs), Piwi-interacting RNA (piRNA), tRNA-derived small RNA (tsRNA) and small rDNA-derived RNA (srRNA).

In Translation

Messenger RNA (mRNA) carries information about a protein sequence to the ribosomes, the protein synthesis factories in the cell. It is coded so that every three nucleotides (a codon) correspond to one amino acid. In eukaryotic cells, once precursor mRNA (pre-mRNA) has been transcribed from DNA, it is processed to mature mRNA. This removes its introns—non-coding sections of the pre-mRNA. The mRNA is then exported from the nucleus to the cytoplasm, where it is bound to ribosomes and translated into its corresponding protein form with the help of tRNA. In prokaryotic cells, which do not have nucleus and cytoplasm compartments, mRNA can bind to ribosomes while it is being transcribed from DNA. After a certain amount of time the message degrades into its component nucleotides with the assistance of ribonucleases.

Transfer RNA (tRNA) is a small RNA chain of about 80 nucleotides that transfers a specific amino acid to a growing polypeptide chain at the ribosomal site of protein synthesis during translation. It has sites for amino acid attachment and an anticodon region for codon recognition that binds to a specific sequence on the messenger RNA chain through hydrogen bonding.

Ribosomal RNA (rRNA) is the catalytic component of the ribosomes. Eukaryotic ribosomes contain four different rRNA molecules: 18S, 5.8S, 28S and 5S rRNA. Three of the rRNA molecules are synthesized in the nucleolus, and one is synthesized elsewhere. In the cytoplasm, ribosomal RNA and protein combine to form a nucleoprotein called a ribosome. The ribosome binds mRNA and carries out protein synthesis. Several ribosomes may be attached to a single mRNA at any time. Nearly all the RNA found in a typical eukaryotic cell is rRNA.

Transfer-messenger RNA (tmRNA) is found in many bacteria and plastids. It tags proteins encoded by mRNAs that lack stop codons for degradation and prevents the ribosome from stalling.

Regulatory RNAs

Several types of RNA can downregulate gene expression by being complementary to a part of an mRNA or a gene's DNA. MicroRNAs (miRNA; 21-22 nt) are found in eukaryotes and act through RNA interference (RNAi), where an effector complex of miRNA and enzymes can cleave complementary mRNA, block the mRNA from being translated, or accelerate its degradation.

While small interfering RNAs (siRNA; 20-25 nt) are often produced by breakdown of viral RNA, there are also endogenous sources of siRNAs. siRNAs act through RNA interference in a fashion similar to miRNAs. Some miRNAs and siRNAs can cause genes they target to be methylated, thereby decreasing or increasing transcription of those genes. Animals have Piwi-interacting RNAs (piRNA; 29-30 nt) that are active in germline cells and are thought to be a defense against transposons and play a role in gametogenesis.

Many prokaryotes have CRISPR RNAs, a regulatory system similar to RNA interference. Antisense RNAs are widespread; most downregulate a gene, but a few are activators of transcription. One way antisense RNA can act is by binding to an mRNA, forming double-stranded RNA that is enzymatically degraded. There are many long noncoding RNAs that regulate genes in eukaryotes, one such RNA is Xist, which coats one X chromosome in female mammals and inactivates it.

An mRNA may contain regulatory elements itself, such as riboswitches, in the 5' untranslated region or 3' untranslated region; these cis-regulatory elements regulate the activity of that mRNA. The untranslated regions can also contain elements that regulate other genes.

In RNA Processing

Many RNAs are involved in modifying other RNAs. Introns are spliced out of pre-mRNA by spliceosomes, which contain several small nuclear RNAs (snRNA), or the introns can be ribozymes that are spliced by themselves. RNA can also be altered by having its nucleotides modified to nucleotides other than A, C, G and U. In eukaryotes, modifications of RNA nucleotides are in general directed by small nucleolar RNAs (snoRNA; 60-300 nt), found in the nucleolus and cajal bodies. snoRNAs associate with enzymes and guide them to a spot on an RNA by basepairing to that RNA. These enzymes then perform the nucleotide modification. rRNAs and tRNAs are extensively modified, but snRNAs and mRNAs can also be the target of base modification. RNA can also be methylated.

RNA Genomes

Like DNA, RNA can carry genetic information. RNA viruses have genomes composed of RNA that encodes a number of proteins. The viral genome is replicated by some of those proteins, while other proteins protect the genome as the virus particle moves to a new host cell. Viroids are another group of pathogens, but they consist only of RNA, do not encode any protein and are replicated by a host plant cell's polymerase.

In Reverse Transcription

Reverse transcribing viruses replicate their genomes by reverse transcribing DNA copies from their RNA; these DNA copies are then transcribed to new RNA. Retrotransposons also spread by copying DNA and RNA from one another, and telomerase contains an RNA that is used as template for building the ends of eukaryotic chromosomes.

Double-stranded RNA

Double-stranded RNA (dsRNA) is RNA with two complementary strands, similar to the DNA found in all cells. dsRNA forms the genetic material of some viruses (double-stranded RNA viruses). Double-stranded RNA such as viral RNA or siRNA can trigger RNA interference in eukaryotes, as well as interferon response in vertebrates.

Circular RNA

Recently, it was shown that there is a single stranded covalently closed, *i.e.* circular form of RNA expressed throughout the animal and plant kingdom. circRNAs are thought to arise via a "back-splice" reaction where the spliceosome joins a downstream donor to an upstream ac-ceptor splice site. So far the function of circRNAs is largely unknown, although for few examples a microRNA sponging activity has been demonstrated.

Key Discoveries in RNA Biology

Research on RNA has led to many important biological discoveries and numerous Nobel Prizes. Nucleic acids were discovered in 1868 by Friedrich Miescher, who called the material 'nuclein' since it was found in the nucleus. It was later discovered that prokaryotic cells, which do not have a nucleus, also contain nucleic acids. The role of RNA in protein synthesis was suspected already in 1939. Severo Ochoa won the 1959 Nobel Prize in Medicine (shared with Arthur Kornberg) after he discovered an enzyme that can synthesize RNA in the laboratory. However, the enzyme discovered by Ochoa (polynucleotide phosphorylase) was later shown to be responsible for RNA degradation, not RNA synthesis. In 1956 Alex Rich and David Davies hybridized two separate strands of RNA to form the first crystal of RNA whose structure could be determined by X-ray crystallography.

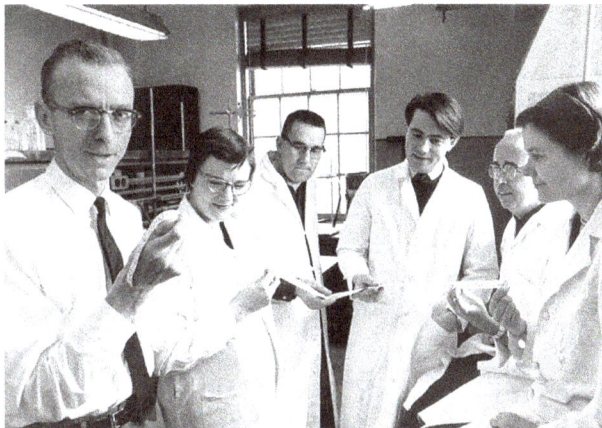

Robert W. Holley, left, poses with his research team.

The sequence of the 77 nucleotides of a yeast tRNA was found by Robert W. Holley in 1965, winning Holley the 1968 Nobel Prize in Medicine (shared with Har Gobind Khorana and Marshall Nirenberg). In 1967, Carl Woese hypothesized that RNA might be catalytic and suggested that the earliest forms of life (self-replicating molecules) could have relied on RNA both to carry genetic information and to catalyze biochemical reactions—an RNA world.

During the early 1970s, retroviruses and reverse transcriptase were discovered, showing for the first time that enzymes could copy RNA into DNA (the opposite of the usual route for transmission of genetic information). For this work, David Baltimore, Renato Dulbecco and Howard Temin were awarded a Nobel Prize in 1975. In 1976, Walter Fiers and his team determined the first complete nucleotide sequence of an RNA virus genome, that of bacteriophage MS2.

In 1977, introns and RNA splicing were discovered in both mammalian viruses and in cellular genes, resulting in a 1993 Nobel to Philip Sharp and Richard Roberts. Catalytic RNA molecules (ribozymes) were discovered in the early 1980s, leading to a 1989 Nobel award to Thomas Cech and Sidney Altman. In 1990, it was found in *Petunia* that introduced genes can silence similar genes of the plant's own, now known to be a result of RNA interference.

At about the same time, 22 nt long RNAs, now called microRNAs, were found to have a role in the development of *C. elegans*. Studies on RNA interference gleaned a Nobel Prize for Andrew Fire and Craig Mello in 2006, and another Nobel was awarded for studies on the transcription of RNA to Roger Kornberg in the same year. The discovery of gene regulatory RNAs has led to attempts to develop drugs made of RNA, such as siRNA, to silence genes.

Evolution

In March 2015, complex DNA and RNA organic compounds of life, including uracil, cytosine and thymine, were reportedly formed in the laboratory under outer space conditions, using starting chemicals, such as pyrimidine, found in meteorites. Pyrimidine, like polycyclic aromatic hydrocarbons (PAHs), the most carbon-rich chemical found in the Universe, may have been formed in red giants or in interstellar dust and gas clouds, according to the scientists.

DNA

Deoxyribonucleic acid is a molecule that carries the genetic instructions used in the growth, development, functioning and reproduction of all known living organisms and many viruses. DNA and RNA are nucleic acids; alongside proteins, lipids and complex carbohydrates (polysaccharides), they are one of the four major types of macromolecules that are essential for all known forms of life. Most DNA molecules consist of two biopolymer strands coiled around each other to form a double helix.

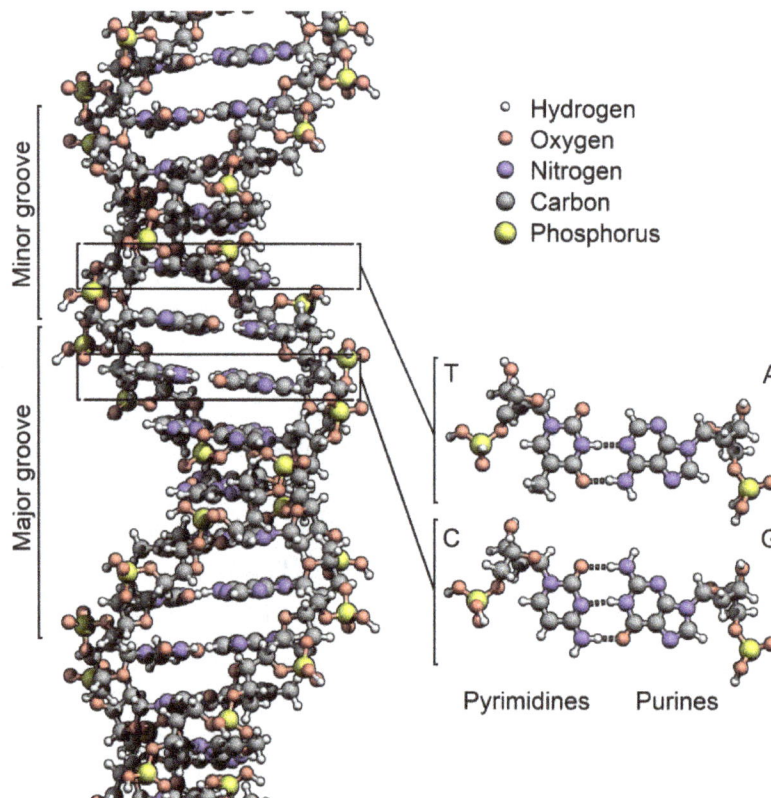

The structure of the DNA double helix. The atoms in the structure are colour-coded by element and the detailed structure of two base pairs are shown in the bottom right.

The two DNA strands are termed polynucleotides since they are composed of simpler monomer units called nucleotides. Each nucleotide is composed of one of four nitrogen-containing nucleobases—either cytosine (C), guanine (G), adenine (A), or thymine (T)—and a sugar called deoxyribose and a phosphate group. The nucleotides are joined to one another in a chain by covalent bonds between the sugar of one nucleotide and the phosphate of the next, resulting in an alternating sugar-phosphate backbone. The nitrogenous bases of the two separate polynucleotide strands are bound together (according to base pairing rules (A with T, and C with G) with hydrogen bonds to make double-stranded DNA. The total amount of related DNA base pairs on Earth is estimated at 5.0×10^{37}, and weighs 50 billion tonnes. In comparison, the total mass of the biosphere has been estimated to be as much as 4 trillion tons of carbon (TtC).

DNA stores biological information. The DNA backbone is resistant to cleavage, and both strands of the double-stranded structure store the same biological information. This information is replicated as and when the two strands separate. A large part of DNA (more than 98% for humans) is non-coding, meaning that these sections do not serve as patterns for protein sequences.

The two strands of DNA run in opposite directions to each other and are thus antiparallel. Attached to each sugar is one of four types of nucleobases (informally, *bases*). It is the sequence of these four nucleobases along the backbone that encodes biological information. RNA strands are created using DNA strands as a template in a process called transcription. Under the genetic code, these RNA strands are translated to specify the sequence of amino acids within proteins in a process called translation.

Within eukaryotic cells, DNA is organized into long structures called chromosomes. During cell division these chromosomes are duplicated in the process of DNA replication, providing each cell its own complete set of chromosomes. Eukaryotic organisms (animals, plants, fungi, and protists) store most of their DNA inside the cell nucleus and some of their DNA in organelles, such as mitochondria or chloroplasts. In contrast, prokaryotes (bacteria and archaea) store their DNA only in the cytoplasm. Within the eukaryotic chromosomes, chromatin proteins such as histones compact and organize DNA. These compact structures guide the interactions between DNA and other proteins, helping control which parts of the DNA are transcribed.

DNA was first isolated by Friedrich Miescher in 1869. Its molecular structure was identified by James Watson and Francis Crick in 1953, whose model-building efforts were guided by X-ray diffraction data acquired by Rosalind Franklin. DNA is used by researchers as a molecular tool to explore physical laws and theories, such as the ergodic theorem and the theory of elasticity. The unique material properties of DNA have made it an attractive molecule for material scientists and engineers interested in micro- and nano-fabrication. Among notable advances in this field are DNA origami and DNA-based hybrid materials.

Properties

DNA is a long polymer made from repeating units called nucleotides. The structure of DNA is non-static, all species comprises two helical chains each coiled round the same axis, and each with a pitch of 34 ångströms (3.4 nanometres) and a radius of 10 ångströms (1.0 nanometre). According to another study, when measured in a particular solution, the DNA chain measured 22 to 26 ångströms wide (2.2 to 2.6 nanometres), and one nucleotide unit measured 3.3 Å (0.33 nm) long. Although each individual repeating unit is very small, DNA polymers can be very large molecules containing millions of nucleotides. For instance, the DNA in the largest human chromosome, chromosome number 1, consists of approximately 220 million base pairs and would be 85 mm long if straightened.

In living organisms DNA does not usually exist as a single molecule, but instead as a pair of molecules that are held tightly together. These two long strands entwine like vines, in the shape of a double helix. The nucleotide contains both a segment of the backbone of the molecule (which holds the chain together) and a nucleobase (which interacts with the other DNA strand in the helix). A nucleobase linked to a sugar is called a nucleoside and a base linked to a sugar and one or more phosphate groups is called a nucleotide. A polymer comprising multiple linked nucleotides (as in DNA) is called a polynucleotide.

The backbone of the DNA strand is made from alternating phosphate and sugar residues. The sugar in DNA is 2-deoxyribose, which is a pentose (five-carbon) sugar. The sugars are joined together by phosphate groups that form phosphodiester bonds between the third and fifth carbon atoms of adjacent sugar rings. These asymmetric bonds mean a strand of DNA has a direction. In a double helix, the direction of the nucleotides in one strand is opposite to their direction in the other strand: the strands are *antiparallel*. The asymmetric ends of DNA strands are said to have a directionality of *five prime* (5′) and *three prime* (3′), with the 5′ end having a terminal phosphate group and the 3′ end a terminal hydroxyl group. One major difference between DNA and RNA is the sugar, with the 2-deoxyribose in DNA being replaced by the alternative pentose sugar ribose in RNA.

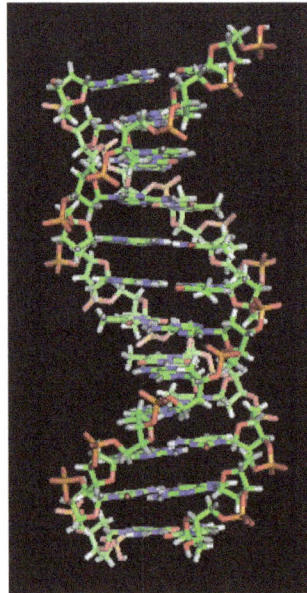

A section of DNA. The bases lie horizontally between the two spiraling strands. (animated version).

The DNA double helix is stabilized primarily by two forces: hydrogen bonds between nucleotides and base-stacking interactions among aromatic nucleobases. In the aqueous environment of the cell, the conjugated π bonds of nucleotide bases align perpendicular to the axis of the DNA molecule, minimizing their interaction with the solvation shell. The four bases found in DNA are adenine (A), cytosine (C), guanine (G) and thymine (T). These four bases are attached to the sugar-phosphate to form the complete nucleotide, as shown for adenosine monophosphate. Adenine pairs with thymine and guanine pairs with cytosine. It was represented by A-T base pairs and G-C base pairs.

Nucleobase Classification

The nucleobases are classified into two types: the purines, A and G, being fused five- and six-membered heterocyclic compounds, and the pyrimidines, the six-membered rings C and T. A fifth pyrimidine nucleobase, uracil (U), usually takes the place of thymine in RNA and differs from thymine by lacking a methyl group on its ring. In addition to RNA and DNA, many artificial nucleic acid analogues have been created to study the properties of nucleic acids, or for use in biotechnology.

Uracil is not usually found in DNA, occurring only as a breakdown product of cytosine. However, in several bacteriophages, *Bacillus subtilis* bacteriophages PBS1 and PBS2 and *Yersinia* bacteriophage piR1-37, thymine has been replaced by uracil. Another phage - Staphylococcal phage S6 - has been identified with a genome where thymine has been replaced by uracil.

Base J (beta-d-glucopyranosyloxymethyluracil), a modified form of uracil, is also found in several organisms: the flagellates *Diplonema* and *Euglena*, and all the kinetoplastid genera. Biosynthesis of J occurs in two steps: in the first step a specific thymidine in DNA is converted into hydroxymethyldeoxyuridine; in the second HOMedU is glycosylated to form J. Proteins that bind specifically to this base have been identified. These proteins appear to be distant relatives of the Tet1 oncogene that is involved in the pathogenesis of acute myeloid leukemia. J appears to act as a termination signal for RNA polymerase II.

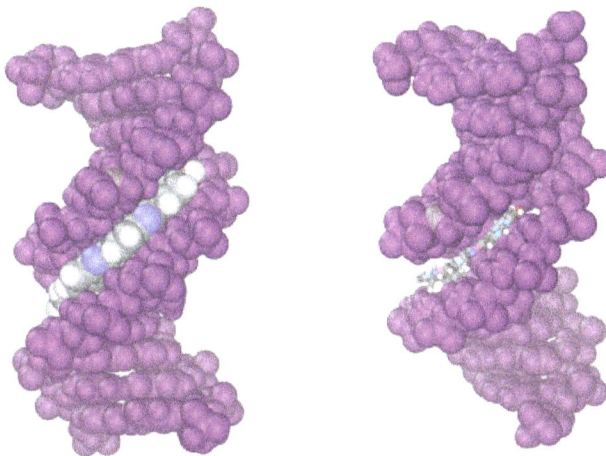

DNA major and minor grooves. The later is a binding site for the Hoechst stain dye 33258.

Grooves

Twin helical strands form the DNA backbone. Another double helix may be found tracing the spaces, or grooves, between the strands. These voids are adjacent to the base pairs and may provide a binding site. As the strands are not symmetrically located with respect to each other, the grooves are unequally sized. One groove, the major groove, is 22 Å wide and the other, the minor groove, is 12 Å wide. The width of the major groove means that the edges of the bases are more accessible in the major groove than in the minor groove. As a result, proteins such as transcription factors that can bind to specific sequences in double-stranded DNA usually make contact with the sides of the bases exposed in the major groove. This situation varies in unusual conformations of DNA within the cell but the major and minor grooves are always named to reflect the differences in size that would be seen if the DNA is twisted back into the ordinary B form.

Base Pairing

In a DNA double helix, each type of nucleobase on one strand bonds with just one type of nucleobase on the other strand. This is called complementary base pairing. Here, purines form hydrogen bonds to pyrimidines, with adenine bonding only to thymine in two hydrogen bonds, and cytosine bonding only to guanine in three hydrogen bonds. This arrangement of two nucleotides binding together across the double helix is called a base pair. As hydrogen bonds are not covalent, they can be broken and rejoined relatively easily. The two strands of DNA in a double helix can thus be pulled apart like a zipper, either by a mechanical force or high temperature. As a result of this base pair complementarity, all the information in the double-stranded sequence of a DNA helix is duplicated on each strand, which is vital in DNA replication. This reversible and specific interaction between complementary base pairs is critical for all the functions of DNA in living organisms.

The two types of base pairs form different numbers of hydrogen bonds, AT forming two hydrogen bonds, and GC forming three hydrogen bonds. DNA with high GC-content is more stable than DNA with low GC-content.

As noted above, most DNA molecules are actually two polymer strands, bound together in a helical fashion by noncovalent bonds; this double stranded structure (dsDNA) is maintained largely by the intrastrand base stacking interactions, which are strongest for G,C stacks. The two strands can come apart – a process known as melting – to form two single-stranded DNA molecules (ssDNA) molecules. Melting occurs at high temperature, low salt and high pH (low pH also melts DNA, but since DNA is unstable due to acid depurination, low pH is rarely used).

The stability of the dsDNA form depends not only on the GC-content (% G,C basepairs) but also on sequence (since stacking is sequence specific) and also length (longer molecules are more stable). The stability can be measured in various ways; a common way is the "melting temperature", which is the temperature at which 50% of the ds molecules are converted to ss molecules; melting temperature is dependent on ionic strength and the concentration of DNA. As a result, it is both the percentage of GC base pairs and the overall length of a DNA double helix that determines the strength of the association between the two strands of DNA. Long DNA helices with a high GC-content have stronger-interacting strands, while short helices with high AT content have weaker-interacting strands. In biology, parts of the DNA double helix that need to separate easily, such as the TATAAT Pribnow box in some promoters, tend to have a high AT content, making the strands easier to pull apart.

In the laboratory, the strength of this interaction can be measured by finding the temperature necessary to break the hydrogen bonds, their melting temperature (also called T_m value). When all the base pairs in a DNA double helix melt, the strands separate and exist in solution as two entirely independent molecules. These single-stranded DNA molecules (*ssDNA*) have no single common shape, but some conformations are more stable than others.

Sense and Antisense

A DNA sequence is called "sense" if its sequence is the same as that of a messenger RNA copy that is translated into protein. The sequence on the opposite strand is called the "antisense" sequence. Both sense and antisense sequences can exist on different parts of the same strand of DNA (i.e.

both strands can contain both sense and antisense sequences). In both prokaryotes and eukaryotes, antisense RNA sequences are produced, but the functions of these RNAs are not entirely clear. One proposal is that antisense RNAs are involved in regulating gene expression through RNA-RNA base pairing.

A few DNA sequences in prokaryotes and eukaryotes, and more in plasmids and viruses, blur the distinction between sense and antisense strands by having overlapping genes. In these cases, some DNA sequences do double duty, encoding one protein when read along one strand, and a second protein when read in the opposite direction along the other strand. In bacteria, this overlap may be involved in the regulation of gene transcription, while in viruses, overlapping genes increase the amount of information that can be encoded within the small viral genome.

Supercoiling

DNA can be twisted like a rope in a process called DNA supercoiling. With DNA in its "relaxed" state, a strand usually circles the axis of the double helix once every 10.4 base pairs, but if the DNA is twisted the strands become more tightly or more loosely wound. If the DNA is twisted in the direction of the helix, this is positive supercoiling, and the bases are held more tightly together. If they are twisted in the opposite direction, this is negative supercoiling, and the bases come apart more easily. In nature, most DNA has slight negative supercoiling that is introduced by enzymes called topoisomerases. These enzymes are also needed to relieve the twisting stresses introduced into DNA strands during processes such as transcription and DNA replication.

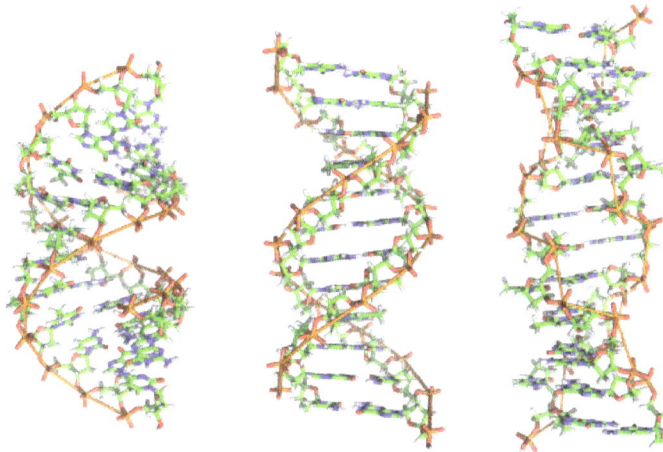

From left to right, the structures of A, B and Z DNA

Alternative DNA Structures

DNA exists in many possible conformations that include A-DNA, B-DNA, and Z-DNA forms, although, only B-DNA and Z-DNA have been directly observed in functional organisms. The conformation that DNA adopts depends on the hydration level, DNA sequence, the amount and direction of supercoiling, chemical modifications of the bases, the type and concentration of metal ions, and the presence of polyamines in solution.

The first published reports of A-DNA X-ray diffraction patterns—and also B-DNA—used analyses based on Patterson transforms that provided only a limited amount of structural information for oriented fibers of DNA. An alternative analysis was then proposed by Wilkins *et al.*, in 1953, for the *in vivo* B-DNA X-ray diffraction-scattering patterns of highly hydrated DNA fibers in terms of squares of Bessel functions. In the same journal, James Watson and Francis Crick presented their molecular modeling analysis of the DNA X-ray diffraction patterns to suggest that the structure was a double-helix.

Although the *B-DNA form* is most common under the conditions found in cells, it is not a well-defined conformation but a family of related DNA conformations that occur at the high hydration levels present in living cells. Their corresponding X-ray diffraction and scattering patterns are characteristic of molecular paracrystals with a significant degree of disorder.

Compared to B-DNA, the A-DNA form is a wider right-handed spiral, with a shallow, wide minor groove and a narrower, deeper major groove. The A form occurs under non-physiological conditions in partly dehydrated samples of DNA, while in the cell it may be produced in hybrid pairings of DNA and RNA strands, and in enzyme-DNA complexes. Segments of DNA where the bases have been chemically modified by methylation may undergo a larger change in conformation and adopt the Z form. Here, the strands turn about the helical axis in a left-handed spiral, the opposite of the more common B form. These unusual structures can be recognized by specific Z-DNA binding proteins and may be involved in the regulation of transcription.

Alternative DNA Chemistry

For many years exobiologists have proposed the existence of a shadow biosphere, a postulated microbial biosphere of Earth that uses radically different biochemical and molecular processes than currently known life. One of the proposals was the existence of lifeforms that use arsenic instead of phosphorus in DNA. A report in 2010 of the possibility in the bacterium GFAJ-1, was announced, though the research was disputed, and evidence suggests the bacterium actively prevents the incorporation of arsenic into the DNA backbone and other biomolecules.

Quadruplex Structures

At the ends of the linear chromosomes are specialized regions of DNA called telomeres. The main function of these regions is to allow the cell to replicate chromosome ends using the enzyme telomerase, as the enzymes that normally replicate DNA cannot copy the extreme 3′ ends of chromosomes. These specialized chromosome caps also help protect the DNA ends, and stop the DNA repair systems in the cell from treating them as damage to be corrected. In human cells, telomeres are usually lengths of single-stranded DNA containing several thousand repeats of a simple TTAGGG sequence.

These guanine-rich sequences may stabilize chromosome ends by forming structures of stacked sets of four-base units, rather than the usual base pairs found in other DNA molecules. Here, four guanine bases form a flat plate and these flat four-base units then stack on top of each other, to form a stable G-quadruplex structure. These structures are stabilized by hydrogen bonding between the edges of the bases and chelation of a metal ion in the centre of each four-base unit. Other structures can also be formed, with the central set of four bases coming from either a single strand folded around the bases, or several different parallel strands, each contributing one base to the central structure.

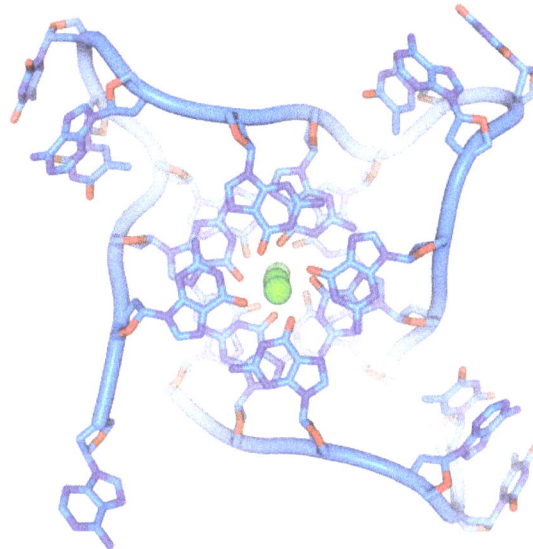

DNA quadruplex formed by telomere repeats. The looped conformation of the DNA backbone is very different from the typical DNA helix. The green spheres in the center represent potassium ions.

In addition to these stacked structures, telomeres also form large loop structures called telomere loops, or T-loops. Here, the single-stranded DNA curls around in a long circle stabilized by telomere-binding proteins. At the very end of the T-loop, the single-stranded telomere DNA is held onto a region of double-stranded DNA by the telomere strand disrupting the double-helical DNA and base pairing to one of the two strands. This triple-stranded structure is called a displacement loop or D-loop.

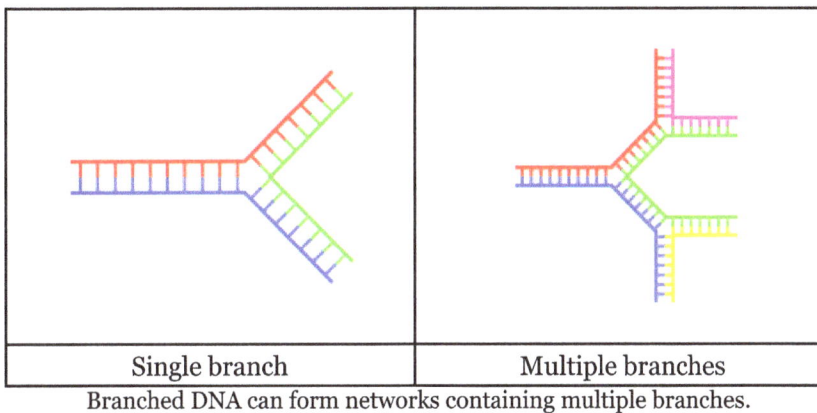

| Single branch | Multiple branches |

Branched DNA can form networks containing multiple branches.

Branched DNA

In DNA, fraying occurs when non-complementary regions exist at the end of an otherwise complementary double-strand of DNA. However, branched DNA can occur if a third strand of DNA is introduced and contains adjoining regions able to hybridize with the frayed regions of the pre-existing double-strand. Although the simplest example of branched DNA involves only three strands of DNA, complexes involving additional strands and multiple branches are also possible. Branched

DNA can be used in nanotechnology to construct geometric shapes, the section on uses in technology below.

Chemical Modifications and Altered DNA packaging

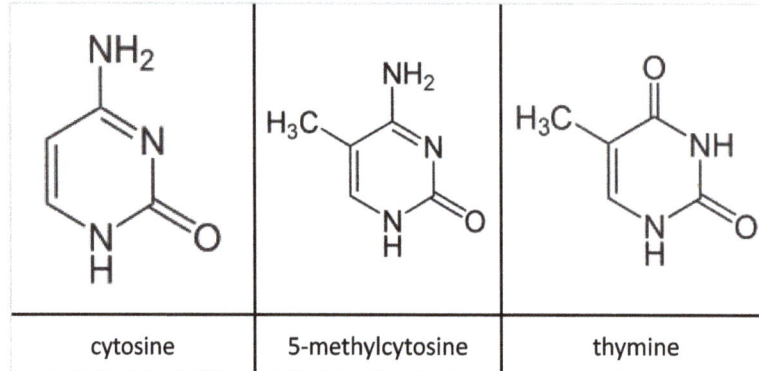

Structure of cytosine with and without the 5-methyl group. Deamination converts 5-methylcytosine into thymine.

Base Modifications and DNA Packaging

The expression of genes is influenced by how the DNA is packaged in chromosomes, in a structure called chromatin. Base modifications can be involved in packaging, with regions that have low or no gene expression usually containing high levels of methylation of cytosine bases. DNA packaging and its influence on gene expression can also occur by covalent modifications of the histone protein core around which DNA is wrapped in the chromatin structure or else by remodeling carried out by chromatin remodeling complexes . There is, further, crosstalk between DNA methylation and histone modification, so they can coordinately affect chromatin and gene expression.

For one example, cytosine methylation, produces 5-methylcytosine, which is important for X-inactivation of chromosomes. The average level of methylation varies between organisms – the worm *Caenorhabditis elegans* lacks cytosine methylation, while vertebrates have higher levels, with up to 1% of their DNA containing 5-methylcytosine. Despite the importance of 5-methylcytosine, it can deaminate to leave a thymine base, so methylated cytosines are particularly prone to mutations. Other base modifications include adenine methylation in bacteria, the presence of 5-hydroxymethylcytosine in the brain, and the glycosylation of uracil to produce the "J-base" in kinetoplastids.

Damage

DNA can be damaged by many sorts of mutagens, which change the DNA sequence. Mutagens include oxidizing agents, alkylating agents and also high-energy electromagnetic radiation such as ultraviolet light and X-rays. The type of DNA damage produced depends on the type of mutagen. For example, UV light can damage DNA by producing thymine dimers, which are cross-links between pyrimidine bases. On the other hand, oxidants such as free radicals or hydrogen peroxide produce multiple forms of damage, including base modifications, particularly of guanosine, and double-strand breaks. A typical human cell contains about 150,000 bases that have suffered oxidative damage. Of these oxidative lesions, the most dangerous are double-strand breaks, as these are difficult to repair and can produce point mutations, insertions, deletions from the DNA sequence, and chromosomal translocations. These mutations can cause cancer. Because of inherent limits in the DNA repair mechanisms, if humans lived long enough, they would all eventually develop cancer. DNA damages that are naturally occurring, due to normal cellular processes that produce reactive oxygen species, the hydrolytic activities of cellular water, etc., also occur frequently. Although most of these damages are repaired, in any cell some DNA damage may remain despite the action of repair processes. These remaining DNA damages accumulate with age in mammalian postmitotic tissues. This accumulation appears to be an important underlying cause of aging.

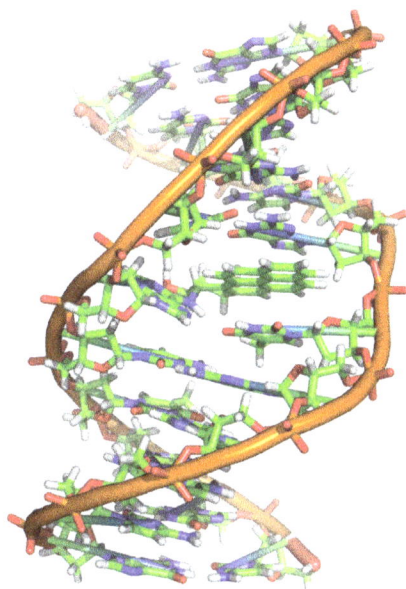

A covalent adduct between a metabolically activated form of benzo[a]pyrene, the major mutagen in tobacco smoke, and DNA

Many mutagens fit into the space between two adjacent base pairs, this is called *intercalation*. Most intercalators are aromatic and planar molecules; examples include ethidium bromide, acridines, daunomycin, and doxorubicin. For an intercalator to fit between base pairs, the bases must separate, distorting the DNA strands by unwinding of the double helix. This inhibits both transcription and DNA replication, causing toxicity and mutations. As a result, DNA intercalators may be carcinogens, and in the case of thalidomide, a teratogen. Others such as benzo[*a*]pyrene diol epoxide and aflatoxin form DNA adducts that induce errors in replication. Nevertheless, due to their ability to inhibit DNA transcription and replication, other similar toxins are also used in chemotherapy to inhibit rapidly growing cancer cells.

Biological Functions

DNA usually occurs as linear chromosomes in eukaryotes, and circular chromosomes in prokaryotes. The set of chromosomes in a cell makes up its genome; the human genome has approximately 3 billion base pairs of DNA arranged into 46 chromosomes. The information carried by DNA is held in the sequence of pieces of DNA called genes. Transmission of genetic information in genes is achieved via complementary base pairing. For example, in transcription, when a cell uses the information in a gene, the DNA sequence is copied into a complementary RNA sequence through the attraction between the DNA and the correct RNA nucleotides. Usually, this RNA copy is then used to make a matching protein sequence in a process called translation, which depends on the same interaction between RNA nucleotides. In alternative fashion, a cell may simply copy its genetic information in a process called DNA replication. The details of these functions are covered in other articles; here the focus is on the interactions between DNA and other molecules that mediate the function of the genome.

Genes and Genomes

Genomic DNA is tightly and orderly packed in the process called DNA condensation, to fit the small available volumes of the cell. In eukaryotes, DNA is located in the cell nucleus, with small amounts in mitochondria and chloroplasts. In prokaryotes, the DNA is held within an irregularly shaped body in the cytoplasm called the nucleoid. The genetic information in a genome is held within genes, and the complete set of this information in an organism is called its genotype. A gene is a unit of heredity and is a region of DNA that influences a particular characteristic in an organism. Genes contain an open reading frame that can be transcribed, and regulatory sequences such as promoters and enhancers, which control transcription of the open reading frame.

In many species, only a small fraction of the total sequence of the genome encodes protein. For example, only about 1.5% of the human genome consists of protein-coding exons, with over 50% of human DNA consisting of non-coding repetitive sequences. The reasons for the presence of so much noncoding DNA in eukaryotic genomes and the extraordinary differences in genome size, or *C-value*, among species represent a long-standing puzzle known as the "C-value enigma". However, some DNA sequences that do not code protein may still encode functional non-coding RNA molecules, which are involved in the regulation of gene expression.

Some noncoding DNA sequences play structural roles in chromosomes. Telomeres and centromeres typically contain few genes, but are important for the function and stability of chromosomes. An abundant form of noncoding DNA in humans are pseudogenes, which are copies of genes that have been disabled by mutation. These sequences are usually just molecular fossils, although they can occasionally serve as raw genetic material for the creation of new genes through the process of gene duplication and divergence.

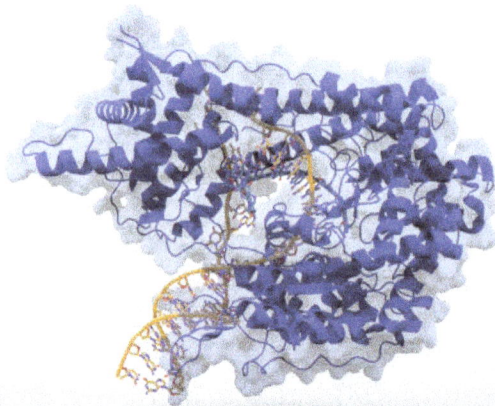

T7 RNA polymerase (blue) producing a mRNA (green) from a DNA template (orange).

Transcription and Translation

A gene is a sequence of DNA that contains genetic information and can influence the phenotype of an organism. Within a gene, the sequence of bases along a DNA strand defines a messenger RNA sequence, which then defines one or more protein sequences. The relationship between the nucleotide sequences of genes and the amino-acid sequences of proteins is determined by the rules of translation, known collectively as the genetic code. The genetic code consists of three-letter 'words' called *codons* formed from a sequence of three nucleotides (e.g. ACT, CAG, TTT).

In transcription, the codons of a gene are copied into messenger RNA by RNA polymerase. This RNA copy is then decoded by a ribosome that reads the RNA sequence by base-pairing the messenger RNA to transfer RNA, which carries amino acids. Since there are 4 bases in 3-letter combinations, there are 64 possible codons (4^3 combinations). These encode the twenty standard amino acids, giving most amino acids more than one possible codon. There are also three 'stop' or 'nonsense' codons signifying the end of the coding region; these are the TAA, TGA, and TAG codons.

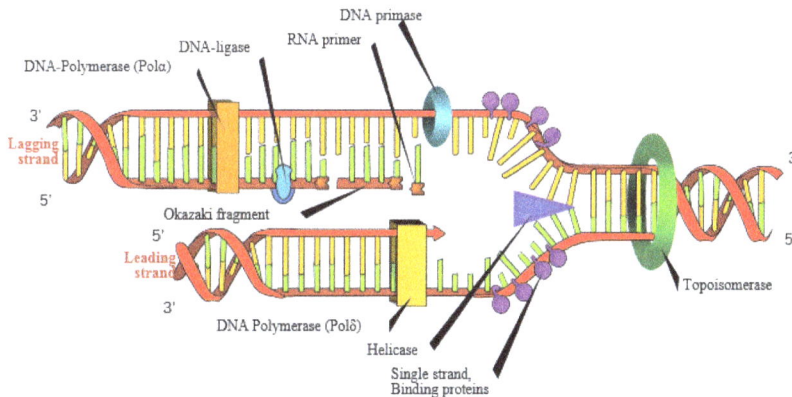

DNA replication. The double helix is unwound by a helicase and topoisomerase. Next, one DNA polymerase produces the leading strand copy. Another DNA polymerase binds to the lagging strand. This enzyme makes discontinuous segments (called Okazaki fragments) before DNA ligase joins them together.

Replication

Cell division is essential for an organism to grow, but, when a cell divides, it must replicate the DNA in its genome so that the two daughter cells have the same genetic information as their parent. The double-stranded structure of DNA provides a simple mechanism for DNA replication. Here, the two strands are separated and then each strand's complementary DNA sequence is recreated by an enzyme called DNA polymerase. This enzyme makes the complementary strand by finding the correct base through complementary base pairing, and bonding it onto the original strand. As DNA polymerases can only extend a DNA strand in a 5' to 3' direction, different mechanisms are used to copy the antiparallel strands of the double helix. In this way, the base on the old strand dictates which base appears on the new strand, and the cell ends up with a perfect copy of its DNA.

Extracellular Nucleic Acids

Naked extracellular DNA (eDNA), most of it released by cell death, is nearly ubiquitous in the environment. Its concentration in soil may be as high as 2 µg/L, and its concentration in natural aquatic environments may be as high at 88 μg/L. Various possible functions have been proposed for eDNA: it may be involved in horizontal gene transfer; it may provide nutrients; and it may act as a buffer to recruit or titrate ions or antibiotics. Extracellular DNA acts as a functional extracellular matrix component in the biofilms of several bacterial species. It may act as a recognition factor to regulate the attachment and dispersal of specific cell types in the biofilm; it may contribute to biofilm formation; and it may contribute to the biofilm's physical strength and resistance to biological stress.

Interactions with Proteins

All the functions of DNA depend on interactions with proteins. These protein interactions can be non-specific, or the protein can bind specifically to a single DNA sequence. Enzymes can also bind to DNA and of these, the polymerases that copy the DNA base sequence in transcription and DNA replication are particularly important.

DNA-binding Proteins

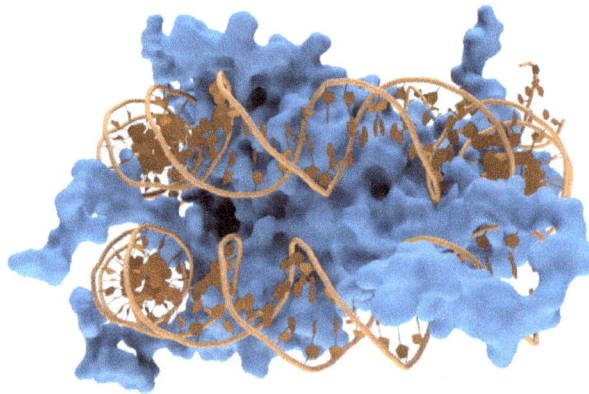

Interaction of DNA (in orange) with histones (in blue). These proteins' basic amino acids bind to the acidic phosphate groups on DNA.

Structural proteins that bind DNA are well-understood examples of non-specific DNA-protein interactions. Within chromosomes, DNA is held in complexes with structural proteins. These proteins organize the DNA into a compact structure called chromatin. In eukaryotes this structure involves DNA binding to a complex of small basic proteins called histones, while in prokaryotes multiple types of proteins are involved. The histones form a disk-shaped complex called a nucleosome, which contains two complete turns of double-stranded DNA wrapped around its surface. These non-specific interactions are formed through basic residues in the histones, making ionic bonds to the acidic sugar-phosphate backbone of the DNA, and are thus largely independent of the base sequence. Chemical modifications of these basic amino acid residues include methylation, phosphorylation and acetylation. These chemical changes alter the strength of the interaction between the DNA and the histones, making the DNA more or less accessible to transcription factors and changing the rate of transcription. Other non-specific DNA-binding proteins in chromatin include the high-mobility group proteins, which bind to bent or distorted DNA. These proteins are important in bending arrays of nucleosomes and arranging them into the larger structures that make up chromosomes.

A distinct group of DNA-binding proteins are the DNA-binding proteins that specifically bind single-stranded DNA. In humans, replication protein A is the best-understood member of this family and is used in processes where the double helix is separated, including DNA replication, recombination and DNA repair. These binding proteins seem to stabilize single-stranded DNA and protect it from forming stem-loops or being degraded by nucleases.

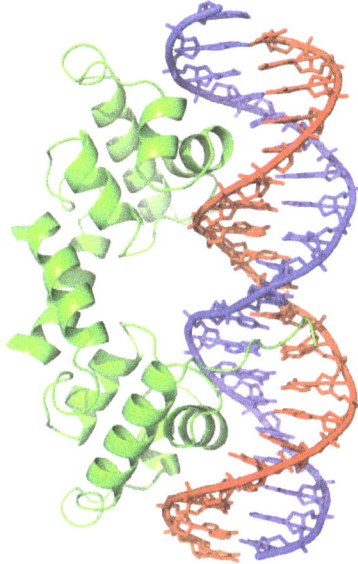

The lambda repressor helix-turn-helix transcription factor bound to its DNA target

In contrast, other proteins have evolved to bind to particular DNA sequences. The most intensively studied of these are the various transcription factors, which are proteins that regulate transcription. Each transcription factor binds to one particular set of DNA sequences and activates or inhibits the transcription of genes that have these sequences close to their promoters. The transcription factors do this in two ways. Firstly, they can bind the RNA polymerase responsible for transcription, either directly or through other mediator proteins; this locates the polymerase at the promoter and allows it to begin transcription. Alternatively, transcription factors can bind enzymes that modify the histones at the promoter. This changes the accessibility of the DNA template to the polymerase.

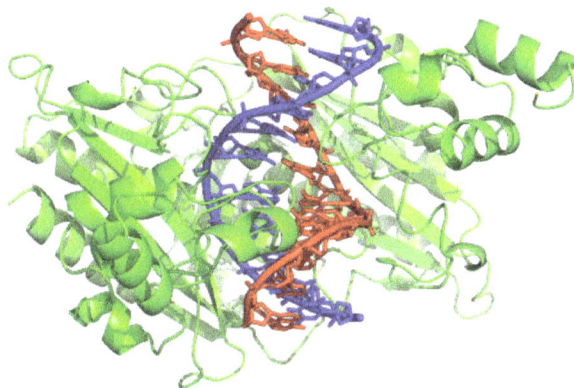

The restriction enzyme EcoRV (green) in a complex with its substrate DNA

As these DNA targets can occur throughout an organism's genome, changes in the activity of one type of transcription factor can affect thousands of genes. Consequently, these proteins are often the targets of the signal transduction processes that control responses to environmental changes or cellular differentiation and development. The specificity of these transcription factors' interactions with DNA come from the proteins making multiple contacts to the edges of the DNA bases, allowing them to "read" the DNA sequence. Most of these base-interactions are made in the major groove, where the bases are most accessible.

DNA-modifying Enzymes

Nucleases and Ligases

Nucleases are enzymes that cut DNA strands by catalyzing the hydrolysis of the phosphodiester bonds. Nucleases that hydrolyse nucleotides from the ends of DNA strands are called exonucleases, while endonucleases cut within strands. The most frequently used nucleases in molecular biology are the restriction endonucleases, which cut DNA at specific sequences. For instance, the EcoRV enzyme shown to the left recognizes the 6-base sequence 5'-GATATC-3' and makes a cut at the vertical line. In nature, these enzymes protect bacteria against phage infection by digesting the phage DNA when it enters the bacterial cell, acting as part of the restriction modification system. In technology, these sequence-specific nucleases are used in molecular cloning and DNA fingerprinting.

Enzymes called DNA ligases can rejoin cut or broken DNA strands. Ligases are particularly important in lagging strand DNA replication, as they join together the short segments of DNA produced at the replication fork into a complete copy of the DNA template. They are also used in DNA repair and genetic recombination.

Topoisomerases and Helicases

Topoisomerases are enzymes with both nuclease and ligase activity. These proteins change the amount of supercoiling in DNA. Some of these enzymes work by cutting the DNA helix and allowing one section to rotate, thereby reducing its level of supercoiling; the enzyme then seals the DNA break. Other types of these enzymes are capable of cutting one DNA helix and then passing a second strand of DNA through this break, before rejoining the helix. Topoisomerases are required for many processes involving DNA, such as DNA replication and transcription.

Helicases are proteins that are a type of molecular motor. They use the chemical energy in nucleoside triphosphates, predominantly adenosine triphosphate (ATP), to break hydrogen bonds between bases and unwind the DNA double helix into single strands. These enzymes are essential for most processes where enzymes need to access the DNA bases.

Polymerases

Polymerases are enzymes that synthesize polynucleotide chains from nucleoside triphosphates. The sequence of their products are created based on existing polynucleotide chains—which are called *templates*. These enzymes function by repeatedly adding a nucleotide to the 3' hydroxyl group at the end of the growing polynucleotide chain. As a consequence, all polymerases work in a

5′ to 3′ direction. In the active site of these enzymes, the incoming nucleoside triphosphate base-pairs to the template: this allows polymerases to accurately synthesize the complementary strand of their template. Polymerases are classified according to the type of template that they use.

In DNA replication, DNA-dependent DNA polymerases make copies of DNA polynucleotide chains. To preserve biological information, it is essential that the sequence of bases in each copy are precisely complementary to the sequence of bases in the template strand. Many DNA polymerases have a proofreading activity. Here, the polymerase recognizes the occasional mistakes in the synthesis reaction by the lack of base pairing between the mismatched nucleotides. If a mismatch is detected, a 3′ to 5′ exonuclease activity is activated and the incorrect base removed. In most organisms, DNA polymerases function in a large complex called the replisome that contains multiple accessory subunits, such as the DNA clamp or helicases.

RNA-dependent DNA polymerases are a specialized class of polymerases that copy the sequence of an RNA strand into DNA. They include reverse transcriptase, which is a viral enzyme involved in the infection of cells by retroviruses, and telomerase, which is required for the replication of telomeres. Telomerase is an unusual polymerase because it contains its own RNA template as part of its structure.

Transcription is carried out by a DNA-dependent RNA polymerase that copies the sequence of a DNA strand into RNA. To begin transcribing a gene, the RNA polymerase binds to a sequence of DNA called a promoter and separates the DNA strands. It then copies the gene sequence into a messenger RNA transcript until it reaches a region of DNA called the terminator, where it halts and detaches from the DNA. As with human DNA-dependent DNA polymerases, RNA polymerase II, the enzyme that transcribes most of the genes in the human genome, operates as part of a large protein complex with multiple regulatory and accessory subunits.

Genetic Recombination

A DNA helix usually does not interact with other segments of DNA, and in human cells the different chromosomes even occupy separate areas in the nucleus called "chromosome territories". This physical separation of different chromosomes is important for the ability of DNA to function as a stable repository for information, as one of the few times chromosomes interact is in chromosomal crossover which occurs during sexual reproduction, when genetic recombination occurs. Chromosomal crossover is when two DNA helices break, swap a section and then rejoin.

Recombination allows chromosomes to exchange genetic information and produces new combinations of genes, which increases the efficiency of natural selection and can be important in the rapid evolution of new proteins. Genetic recombination can also be involved in DNA repair, particularly in the cell's response to double-strand breaks.

The most common form of chromosomal crossover is homologous recombination, where the two chromosomes involved share very similar sequences. Non-homologous recombination can be damaging to cells, as it can produce chromosomal translocations and genetic abnormalities. The recombination reaction is catalyzed by enzymes known as recombinases, such as RAD51. The first step in recombination is a double-stranded break caused by either an endonuclease or damage to the DNA. A series of steps catalyzed in part by the recombinase then leads to joining of the two helices by at least one Holliday junction, in which a segment of a single strand in each helix is annealed to the complementary strand in the other helix. The Holliday junction is a tetrahedral junction structure that can be moved along the pair of chromosomes, swapping one strand for another. The recombination reaction is then halted by cleavage of the junction and re-ligation of the released DNA.

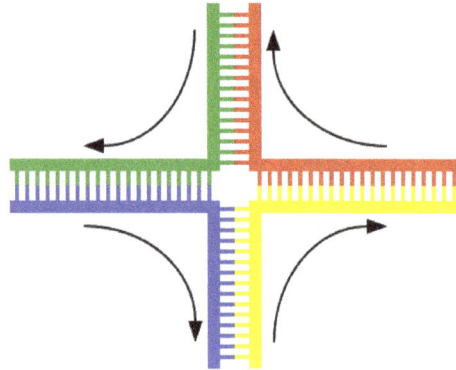

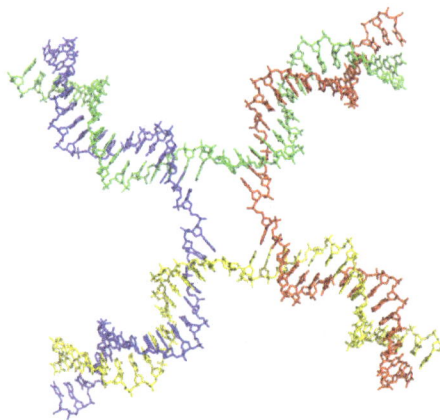

Structure of the Holliday junction intermediate in genetic recombination. The four separate DNA strands are coloured red, blue, green and yellow.

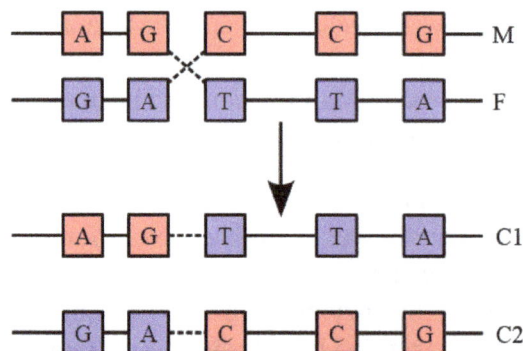

Recombination involves the breaking and rejoining of two chromosomes (M and F) to produce two rearranged chromosomes (C1 and C2).

Evolution

DNA contains the genetic information that allows all modern living things to function, grow and reproduce. However, it is unclear how long in the 4-billion-year history of life DNA has performed this function, as it has been proposed that the earliest forms of life may have used RNA as their genetic material. RNA may have acted as the central part of early cell metabolism as it can both transmit genetic information and carry out catalysis as part of ribozymes. This ancient RNA world where nucleic acid would have been used for both catalysis and genetics may have influenced the evolution of the current genetic code based on four nucleotide bases. This would occur, since the number of different bases in such an organism is a trade-off between a small number of bases increasing replication accuracy and a large number of bases increasing the catalytic efficiency of ribozymes. However, there is no direct evidence of ancient genetic systems, as recovery of DNA from most fossils is impossible because DNA survives in the environment for less than one million years, and slowly degrades into short fragments in solution. Claims for older DNA have been made, most notably a report of the isolation of a viable bacterium from a salt crystal 250 million years old, but these claims are controversial.

Building blocks of DNA (adenine, guanine and related organic molecules) may have been formed extraterrestrially in outer space. Complex DNA and RNA organic compounds of life, including uracil, cytosine, and thymine, have also been formed in the laboratory under conditions mimicking those found in outer space, using starting chemicals, such as pyrimidine, found in meteorites. Pyrimidine, like polycyclic aromatic hydrocarbons (PAHs), the most carbon-rich chemical found in the universe, may have been formed in red giants or in interstellar cosmic dust and gas clouds.

Uses in Technology

Genetic Engineering

Methods have been developed to purify DNA from organisms, such as phenol-chloroform extraction, and to manipulate it in the laboratory, such as restriction digests and the polymerase chain reaction. Modern biology and biochemistry make intensive use of these techniques in recombinant DNA technology. Recombinant DNA is a man-made DNA sequence that has been assembled from other DNA sequences. They can be transformed into organisms in the form of plasmids or in the appropriate format, by using a viral vector. The genetically modified organisms produced can be used to produce products such as recombinant proteins, used in medical research, or be grown in agriculture.

DNA Profiling

Forensic scientists can use DNA in blood, semen, skin, saliva or hair found at a crime scene to identify a matching DNA of an individual, such as a perpetrator. This process is formally termed DNA profiling, but may also be called "genetic fingerprinting". In DNA profiling, the lengths of variable sections of repetitive DNA, such as short tandem repeats and minisatellites, are compared between people. This method is usually an extremely reliable technique for identifying a matching DNA. However, identification can be complicated if the scene is contaminated with DNA from several people. DNA profiling was developed in 1984 by British geneticist Sir Alec Jeffreys, and first used in forensic science to convict Colin Pitchfork in the 1988 Enderby murders case.

The development of forensic science, and the ability to now obtain genetic matching on minute samples of blood, skin, saliva, or hair has led to re-examining many cases. Evidence can now be uncovered that was scientifically impossible at the time of the original examination. Combined with the removal of the double jeopardy law in some places, this can allow cases to be reopened where prior trials have failed to produce sufficient evidence to convince a jury. People charged with serious crimes may be required to provide a sample of DNA for matching purposes. The most obvious defence to DNA matches obtained forensically is to claim that cross-contamination of evidence has occurred. This has resulted in meticulous strict handling procedures with new cases of serious crime. DNA profiling is also used successfully to positively identify victims of mass casualty incidents, bodies or body parts in serious accidents, and individual victims in mass war graves, via matching to family members.

DNA profiling is also used in DNA paternity testing to determine if someone is the biological parent or grandparent of a child with the probability of parentage is typically 99.99% when the alleged parent is biologically related to the child. Normal DNA sequencing methods happen after birth but there are new methods to test paternity while a mother is still pregnant.

DNA Enzymes or Catalytic DNA

Deoxyribozymes, also called DNAzymes or catalytic DNA are first discovered in 1994. They are mostly single stranded DNA sequences isolated from a large pool of random DNA sequences through a combinatorial approach called in vitro selection or systematic evolution of ligands by exponential enrichment (SELEX). DNAzymes catalyze variety of chemical reactions including RNA-DNA cleavage, RNA-DNA ligation, amino acids phosphorylation-dephosphorylation, carbon-carbon bond formation, and etc. DNAzymes can enhance catalytic rate of chemical reactions up to 100,000,000,000-fold over the uncatalyzed reaction. The most extensively studied class of DNAzymes are RNA-cleaving types which have been used to detect different metal ions and designing therapeutic agents. Several metal-specific DNAzymes have been reported including the GR-5 DNAzyme (lead-specific), the CA1-3 DNAzymes (copper-specific), the 39E DNAzyme (uranyl-specific) and the NaA43 DNAzyme (sodium-specific). The NaA43 DNAzyme, which is reported to be more than 10,000-fold selective for sodium over other metal ions, was used to make a real-time sodium sensor in living cells.

Bioinformatics

Bioinformatics involves the development of techniques to store, data mine, search and manipulate biological data, including DNA nucleic acid sequence data. These have led to widely applied advances in computer science, especially string searching algorithms, machine learning and database theory. String searching or matching algorithms, which find an occurrence of a sequence of letters inside a larger sequence of letters, were developed to search for specific sequences of nucleotides. The DNA sequence may be aligned with other DNA sequences to identify homologous sequences and locate the specific mutations that make them distinct. These techniques, especially multiple sequence alignment, are used in studying phylogenetic relationships and protein function. Data sets representing entire genomes' worth of DNA sequences, such as those produced by the Human Genome Project, are difficult to use without the annotations that identify the locations of genes and regulatory elements on each chromosome. Regions of DNA sequence that have the characteristic patterns associated with protein- or RNA-coding genes can be identified by gene finding algorithms, which allow researchers to predict the presence of particular gene products and their possible functions in an organism even before they have been isolated experimentally. Entire genomes may also be compared, which can shed light on the evolutionary history of particular organism and permit the examination of complex evolutionary events.

DNA Nanotechnology

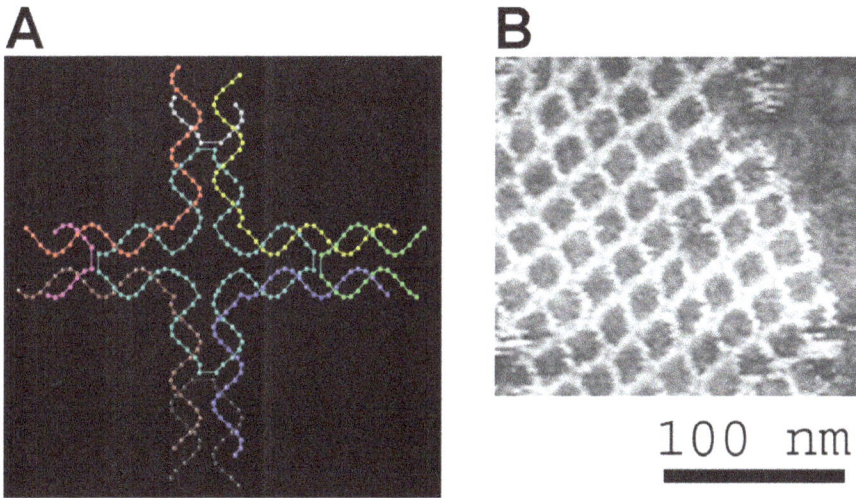

The DNA structure at left (schematic shown) will self-assemble into the structure visualized by atomic force microscopy at right. DNA nanotechnology is the field that seeks to design nanoscale structures using the molecular recognition properties of DNA molecules. Image from Strong, 2004.

DNA nanotechnology uses the unique molecular recognition properties of DNA and other nucleic acids to create self-assembling branched DNA complexes with useful properties. DNA is thus used as a structural material rather than as a carrier of biological information. This has led to the creation of two-dimensional periodic lattices (both tile-based and using the *DNA origami* method) and three-dimensional structures in the shapes of polyhedra. Nanomechanical devices and algorithmic self-assembly have also been demonstrated, and these DNA structures have been used to template the arrangement of other molecules such as gold nanoparticles and streptavidin proteins.

History and Anthropology

Because DNA collects mutations over time, which are then inherited, it contains historical information, and, by comparing DNA sequences, geneticists can infer the evolutionary history of organisms, their phylogeny. This field of phylogenetics is a powerful tool in evolutionary biology. If DNA sequences within a species are compared, population geneticists can learn the history of particular populations. This can be used in studies ranging from ecological genetics to anthropology; For example, DNA evidence is being used to try to identify the Ten Lost Tribes of Israel.

Information Storage

In a paper published in *Nature* in January 2013, scientists from the European Bioinformatics Institute and Agilent Technologies proposed a mechanism to use DNA's ability to code information as a means of digital data storage. The group was able to encode 739 kilobytes of data into DNA code, synthesize the actual DNA, then sequence the DNA and decode the information back to its original form, with a reported 100% accuracy. The encoded information consisted of text files and audio files. A prior experiment was published in August 2012. It was conducted by researchers at Harvard University, where the text of a 54,000-word book was encoded in DNA.

History of DNA Research

James Watson and Francis Crick (right), co-originators of the double-helix model, with Maclyn McCarty (left).

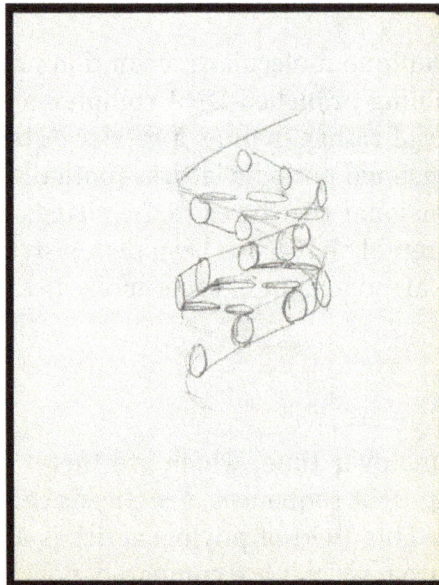

Pencil sketch of the DNA double helix by Francis Crick in 1953

DNA was first isolated by the Swiss physician Friedrich Miescher who, in 1869, discovered a microscopic substance in the pus of discarded surgical bandages. As it resided in the nuclei of cells, he called it "nuclein". In 1878, Albrecht Kossel isolated the non-protein component of "nuclein", nucleic acid, and later isolated its five primary nucleobases. In 1919, Phoebus Levene identified the base, sugar and phosphate nucleotide unit. Levene suggested that DNA consisted of a string of nucleotide units linked together through the phosphate groups. Levene thought the chain was short and the bases repeated in a fixed order. In 1937, William Astbury produced the first X-ray diffraction patterns that showed that DNA had a regular structure.

In 1927, Nikolai Koltsov proposed that inherited traits would be inherited via a "giant hereditary molecule" made up of "two mirror strands that would replicate in a semi-conservative fashion using each strand as a template". In 1928, Frederick Griffith in his experiment discovered that traits of the "smooth" form of *Pneumococcus* could be transferred to the "rough" form of the same bacteria by mixing killed "smooth" bacteria with the live "rough" form. This system provided the first clear suggestion that DNA carries genetic information—the Avery–MacLeod–McCarty experiment—when Oswald Avery, along with coworkers Colin MacLeod and Maclyn McCarty, identified DNA as the transforming principle in 1943. DNA's role in heredity was confirmed in 1952, when Alfred Hershey and Martha Chase in the Hershey–Chase experiment showed that DNA is the genetic material of the T2 phage.

In 1953, James Watson and Francis Crick suggested what is now accepted as the first correct double-helix model of DNA structure in the journal *Nature*. Their double-helix, molecular model of DNA was then based on one X-ray diffraction image (labeled as "Photo 51") taken by Rosalind Franklin and Raymond Gosling in May 1952, and the information that the DNA bases are paired.

Experimental evidence supporting the Watson and Crick model was published in a series of five articles in the same issue of *Nature*. Of these, Franklin and Gosling's paper was the first publication of their own X-ray diffraction data and original analysis method that partly supported the Watson and Crick model; this issue also contained an article on DNA structure by Maurice Wilkins and two of his colleagues, whose analysis and *in vivo* B-DNA X-ray patterns also supported the presence *in vivo* of the double-helical DNA configurations as proposed by Crick and Watson for their double-helix molecular model of DNA in the prior two pages of *Nature*. In 1962, after Franklin's death, Watson, Crick, and Wilkins jointly received the Nobel Prize in Physiology or Medicine. Nobel Prizes are awarded only to living recipients. A debate continues about who should receive credit for the discovery.

In an influential presentation in 1957, Crick laid out the central dogma of molecular biology, which foretold the relationship between DNA, RNA, and proteins, and articulated the "adaptor hypothesis". Final confirmation of the replication mechanism that was implied by the double-helical structure followed in 1958 through the Meselson–Stahl experiment. Further work by Crick and coworkers showed that the genetic code was based on non-overlapping triplets of bases, called codons, allowing Har Gobind Khorana, Robert W. Holley and Marshall Warren Nirenberg to decipher the genetic code. These findings represent the birth of molecular biology.

DNA Sequencing

DNA sequencing is the process of determining the precise order of nucleotides within a DNA molecule. It includes any method or technology that is used to determine the order of the four bases—adenine, guanine, cytosine, and thymine—in a strand of DNA. The advent of rapid DNA sequencing methods has greatly accelerated biological and medical research and discovery.

Knowledge of DNA sequences has become indispensable for basic biological research, and in numerous applied fields such as medical diagnosis, biotechnology, forensic biology, virology and biological systematics. The rapid speed of sequencing attained with modern DNA sequencing technology has been instrumental in the sequencing of complete DNA sequences, or genomes of numerous types and species of life, including the human genome and other complete DNA sequences of many animal, plant, and microbial species.

An example of the results of automated chain-termination DNA sequencing.

The first DNA sequences were obtained in the early 1970s by academic researchers using laborious methods based on two-dimensional chromatography. Following the development of fluorescence-based sequencing methods with a DNA sequencer, DNA sequencing has become easier and orders of magnitude faster.

Applications

DNA sequencing may be used to determine the sequence of individual genes, larger genetic regions (i.e. clusters of genes or operons), full chromosomes or entire genomes, of any organism. DNA sequencing is also the most efficient way to sequence RNA or proteins (via their open reading frames). In fact, DNA sequencing has become a key technology in many areas of biology and other sciences such as medicine, forensics, or anthropology. Some applications include:

Molecular Biology

Sequencing is used in molecular biology to study genomes and the proteins they encode. Information obtained using sequencing allows researchers to identify changes in genes, associations with diseases and phenotypes, and identify potential drug targets.

Evolutionary Biology

Since DNA is an informative macromolecule in terms of transmission from one generation to another, DNA sequencing is used in evolutionary biology to study how different organisms are related and how they evolved.

Metagenomics

The field of metagenomics involves identification of organisms present in a body of water, sewage, dirt, debris filtered from the air, or swab samples from organisms. Knowing which organisms are present in a particular environment is critical to research in ecology, epidemiology, microbiology, and other fields. Sequencing enables researchers to determine which types of microbes may be present in a microbiome, for example.

Medicine

Medical technicians may sequence genes (or, theoretically, full genomes) from patients to determine if there is risk of genetic diseases. This is a form of genetic testing, though some genetic tests may not involve DNA sequencing.

Forensics

DNA sequencing may be used along with DNA profiling methods for forensic identification and paternity testing.

The Four Canonical Bases

The canonical structure of DNA has four bases: thymine (T), adenine (A), cytosine (C), and guanine (G). DNA sequencing is the determination of the physical order of these bases in a molecule of DNA. However, there are many other bases that may be present in a molecule. In some viruses (specifically, bacteriophage), cytosine may be replaced by hydroxy methyl or hydroxy methyl glucose cytosine. In mammalian DNA, variant bases with methyl groups or phosphosulfate may be found. Depending on the sequencing technique, a particular modification, e.g., the 5mC (5 methyl cytosine) common in humans, may or may not be detected.

History

Deoxyribonucleic acid (DNA) was first discovered and isolated by Friedrich Miescher in 1869, but it remained understudied for many decades because proteins, rather than DNA, were thought to hold the genetic blueprint to life. This situation changed after 1944 as a result of some experiments by Oswald Avery, Colin MacLeod, and Maclyn McCarty demonstrating that purified DNA could change one strain of bacteria into another. This was the first time that DNA was shown capable of transforming the properties of cells.

In 1953 James Watson and Francis Crick put forward their double-helix model of DNA, based on crystallized X-ray structures being studied by Rosalind Franklin. According to the model, DNA is composed of two strands of nucleotides coiled around each other, linked together by hydrogen bonds and running in opposite directions. Each strand is composed of four complementary nucleotides – adenine (A), cytosine (C), guanine (G) and thymine (T) – with an A on one strand always paired with T on the other, and C always paired with G. They proposed such a structure allowed each strand to be used to reconstruct the other, an idea central to the passing on of hereditary information between generations.

Frederick Sanger, a pioneer of sequencing. Sanger is one of the few scientists who was awarded two Nobel prizes, one for the sequencing of proteins, and the other for the sequencing of DNA.

The foundation for sequencing proteins was first laid by the work of Fred Sanger who by 1955 had completed the sequence of all the amino acids in insulin, a small protein secreted by the pancreas. This provided the first conclusive evidence that proteins were chemical entities with a specific molecular pattern rather than a random mixture of material suspended in fluid. Sanger's success in sequencing insulin greatly electrified x-ray crystallographers, including Watson and Crick who by now were trying to understand how DNA directed the formation of proteins within a cell. Soon after attending a series of lectures given by Fred Sanger in October 1954, Crick began to develop a theory which argued that the arrangement of nucleotides in DNA determined the sequence of amino acids in proteins which in turn helped determine the function of a protein. He published this theory in 1958.

RNA Sequencing

RNA sequencing was one of the earliest forms of nucleotide sequencing. The major landmark of RNA sequencing is the sequence of the first complete gene and the complete genome of Bacteriophage MS2, identified and published by Walter Fiers and his coworkers at the University of Ghent (Ghent, Belgium), in 1972 and 1976. Traditional RNA sequencing methods require the creation of a cDNA molecule which must be sequenced. In August 2016, scientists at Oxford Nanopore Technologies published a pre-print indicating that direct RNA sequencing could be performed in real time using nanopore technology.

Early DNA Sequencing Methods

The first method for determining DNA sequences involved a location-specific primer extension strategy established by Ray Wu at Cornell University in 1970. DNA polymerase catalysis and specific nucleotide labeling, both of which figure prominently in current sequencing schemes, were used to sequence the cohesive ends of lambda phage DNA. Between 1970 and 1973, Wu, R Padmanabhan and colleagues demonstrated that this method can be employed to determine any DNA sequence using synthetic location-specific primers. Frederick Sanger then adopted this primer-extension strategy to develop more rapid DNA sequencing methods at the MRC Centre, Cambridge, UK and published a method for "DNA sequencing with chain-terminating inhibitors" in 1977. Walter Gilbert and Allan Maxam at Harvard also developed sequencing methods, including one for "DNA sequencing by chemical degradation". In 1973, Gilbert and Maxam reported the sequence of 24 basepairs using a method known as wandering-spot analysis. Advancements in sequencing were aided by the concurrent development of recombinant DNA technology, allowing DNA samples to be isolated from sources other than viruses.

Sequencing of Full Genomes

The first full DNA genome to be sequenced was that of bacteriophage φX174 in 1977. Medical Research Council scientists deciphered the complete DNA sequence of the Epstein-Barr virus in 1984, finding it contained 172,282 nucleotides. Completion of the sequence marked a significant turning point in DNA sequencing because it was achieved with no prior genetic profile knowledge of the virus.

A non-radioactive method for transferring the DNA molecules of sequencing reaction mixtures onto an immobilizing matrix during electrophoresis was developed by Pohl and co-workers in the early 1980s. Followed by the commercialization of the DNA sequencer "Direct-Blotting-Electrophoresis-System GATC 1500" by GATC Biotech, which was intensively used in the framework of the EU genome-sequencing programme, the complete DNA sequence of the yeast *Saccharomyces cerevisiae* chromosome II. Leroy E. Hood's laboratory at the California Institute of Technology announced the first semi-automated DNA sequencing machine in 1986. This was followed by Applied Biosystems' marketing of the first fully automated sequencing machine, the ABI 370, in 1987 and by Dupont's Genesis 2000 which used a novel fluorescent labeling technique enabling all four dideoxynucleotides to be identified in a single lane. By 1990, the U.S. National Institutes of Health (NIH) had begun large-scale sequencing trials on *Mycoplasma capricolum*, *Escherichia coli*, *Caenorhabditis elegans*, and *Saccharomyces cerevisiae* at a cost of US$0.75 per base. Meanwhile, sequencing of human cDNA sequences called expressed sequence tags began in Craig Venter's lab, an attempt to capture the coding fraction of the human genome. In 1995, Venter, Hamilton Smith, and colleagues at The Institute for Genomic Research (TIGR) published the first complete genome of a free-living organism, the bacterium *Haemophilus influenzae*. The circular chromosome contains 1,830,137 bases and its publication in the journal Science marked the first published use of whole-genome shotgun sequencing, eliminating the need for initial mapping efforts.

By 2001, shotgun sequencing methods had been used to produce a draft sequence of the human genome.

Next-generation Sequencing (NGS) Methods

Several new methods for DNA sequencing were developed in the mid to late 1990s and were implemented in commercial DNA sequencers by the year 2000.

On October 26, 1990, Roger Tsien, Pepi Ross, Margaret Fahnestock and Allan J Johnston filed a patent describing stepwise ("base-by-base") sequencing with removable 3' blockers on DNA arrays (blots and single DNA molecules). In 1996, Pål Nyrén and his student Mostafa Ronaghi at the Royal Institute of Technology in Stockholm published their method of pyrosequencing.

On April 1, 1997, Pascal Mayer and Laurent Farinelli submitted patents to the World Intellectual Property Organization describing DNA colony sequencing. The DNA sample preparation and random surface-PCR arraying methods described in this patent, coupled to Roger Tsien et al.'s "base-by-base" sequencing method, is now implemented in Illumina's Hi-Seq genome sequencers.

Lynx Therapeutics published and marketed "Massively parallel signature sequencing", or MPSS, in 2000. This method incorporated a parallelized, adapter/ligation-mediated, bead-based sequencing technology and served as the first commercially available "next-generation" sequencing method, though no DNA sequencers were sold to independent laboratories.

In 2004, 454 Life Sciences marketed a parallelized version of pyrosequencing. The first version of their machine reduced sequencing costs 6-fold compared to automated Sanger sequencing, and was the second of the new generation of sequencing technologies, after MPSS.

The large quantities of data produced by DNA sequencing have also required development of new methods and programs for sequence analysis. Phil Green and Brent Ewing of the University of Washington described their phred quality score for sequencer data analysis in 1998.

Basic Methods

Maxam-gilbert Sequencing

Allan Maxam and Walter Gilbert published a DNA sequencing method in 1977 based on chemical modification of DNA and subsequent cleavage at specific bases. Also known as chemical sequencing, this method allowed purified samples of double-stranded DNA to be used without further cloning. This method's use of radioactive labeling and its technical complexity discouraged extensive use after refinements in the Sanger methods had been made.

Maxam-Gilbert sequencing requires radioactive labeling at one 5' end of the DNA and purification of the DNA fragment to be sequenced. Chemical treatment then generates breaks at a small proportion of one or two of the four nucleotide bases in each of four reactions (G, A+G, C, C+T). The concentration of the modifying chemicals is controlled to introduce on average one modification per DNA molecule. Thus a series of labeled fragments is generated, from the radiolabeled end to the first "cut" site in each molecule. The fragments in the four reactions are electrophoresed side by side in denaturing acrylamide gels for size separation. To visualize the fragments, the gel is exposed to X-ray film for autoradiography, yielding a series of dark bands each corresponding to a radiolabeled DNA fragment, from which the sequence may be inferred.

Chain-termination Methods

The chain-termination method developed by Frederick Sanger and coworkers in 1977 soon became the method of choice, owing to its relative ease and reliability. When invented, the chain-terminator method used fewer toxic chemicals and lower amounts of radioactivity than the Maxam and Gilbert method. Because of its comparative ease, the Sanger method was soon automated and was the method used in the first generation of DNA sequencers.

Sanger sequencing is the method which prevailed from the 1980s until the mid-2000s. Over that period, great advances were made in the technique, such as fluorescent labelling, capillary electrophoresis, and general automation. These developments allowed much more efficient sequencing, leading to lower costs. The Sanger method, in mass production form, is the technology which produced the first human genome in 2001, ushering in the age of genomics. However, later in the decade, radically different approaches reached the market, bringing the cost per genome down from $100 million in 2001 to $10,000 in 2011.

Advanced Methods and De Novo Sequencing

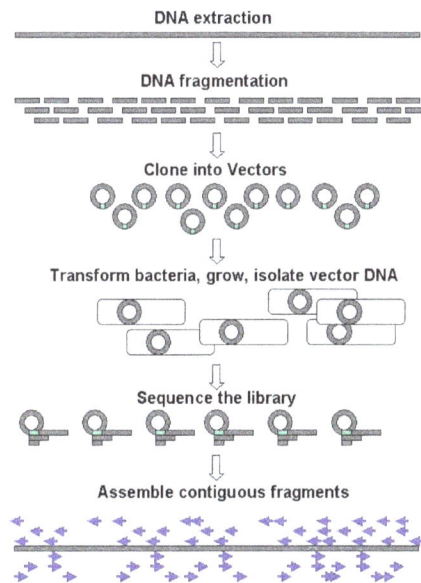

Genomic DNA is fragmented into random pieces and cloned as a bacterial library. DNA from individual bacterial clones is sequenced and the sequence is assembled by using overlapping DNA regions.(click to expand)

Large-scale sequencing often aims at sequencing very long DNA pieces, such as whole chromosomes, although large-scale sequencing can also be used to generate very large numbers of short sequences, such as found in phage display. For longer targets such as chromosomes, common approaches consist of cutting (with restriction enzymes) or shearing (with mechanical forces) large DNA fragments into shorter DNA fragments. The fragmented DNA may then be cloned into a DNA vector and amplified in a bacterial host such as *Escherichia coli*. Short DNA fragments purified from individual bacterial colonies are individually sequenced and assembled electronically into one long, contiguous sequence. Studies have shown that adding a size selection step to collect DNA fragments of uniform size can improve sequencing efficiency and accuracy of the genome assembly. In these studies, automated sizing has proven to be more reproducible and precise than manual gel sizing.

The term "*de novo* sequencing" specifically refers to methods used to determine the sequence of DNA with no previously known sequence. *De novo* translates from Latin as "from the beginning". Gaps in the assembled sequence may be filled by primer walking. The different strategies have different tradeoffs in speed and accuracy; shotgun methods are often used for sequencing large genomes, but its assembly is complex and difficult, particularly with sequence repeats often causing gaps in genome assembly.

Most sequencing approaches use an *in vitro* cloning step to amplify individual DNA molecules, because their molecular detection methods are not sensitive enough for single molecule sequencing. Emulsion PCR isolates individual DNA molecules along with primer-coated beads in aqueous droplets within an oil phase. A polymerase chain reaction (PCR) then coats each bead with clonal copies of the DNA molecule followed by immobilization for later sequencing. Emulsion PCR is used in the methods developed by Marguilis et al. (commercialized by 454 Life Sciences), Shendure and Porreca et al. (also known as "Polony sequencing") and SOLiD sequencing, (developed by Agencourt, later Applied Biosystems, now Life Technologies).

Shotgun Sequencing

Shotgun sequencing is a sequencing method designed for analysis of DNA sequences longer than 1000 base pairs, up to and including entire chromosomes. This method requires the target DNA to be broken into random fragments. After sequencing individual fragments, the sequences can be reassembled on the basis of their overlapping regions.

Bridge PCR

Another method for *in vitro* clonal amplification is bridge PCR, in which fragments are amplified upon primers attached to a solid surface and form "DNA colonies" or "DNA clusters". This method is used in the Illumina Genome Analyzer sequencers. Single-molecule methods, such as that developed by Stephen Quake's laboratory (later commercialized by Helicos) are an exception: they use bright fluorophores and laser excitation to detect base addition events from individual DNA molecules fixed to a surface, eliminating the need for molecular amplification.

Next-generation Methods

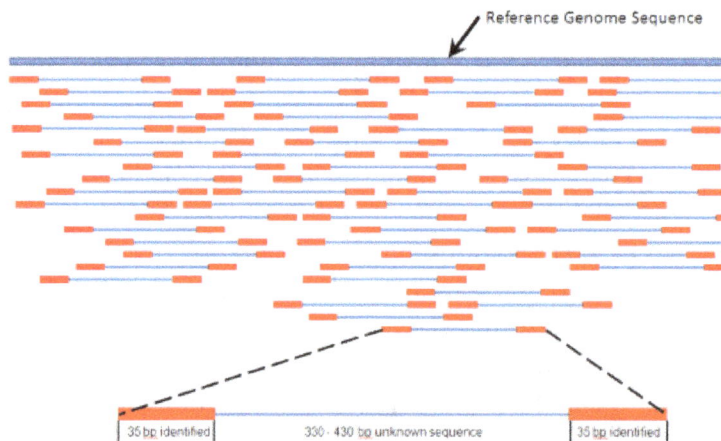

Multiple, fragmented sequence reads must be assembled together on the basis of their overlapping areas.

Next-generation sequencing applies to genome sequencing, genome resequencing, transcriptome profiling (RNA-Seq), DNA-protein interactions (ChIP-sequencing), and epigenome characterization. Resequencing is necessary, because the genome of a single individual of a species will not indicate all of the genome variations among other individuals of the same species.

The high demand for low-cost sequencing has driven the development of high-throughput sequencing (or next-generation sequencing) technologies that parallelize the sequencing process, producing thousands or millions of sequences concurrently. High-throughput sequencing technologies are intended to lower the cost of DNA sequencing beyond what is possible with standard dye-terminator methods. In ultra-high-throughput sequencing as many as 500,000 sequencing-by-synthesis operations may be run in parallel.

Comparison of next-generation sequencing methods							
Method	**Read length**	**Accuracy (single read not consensus)**	**Reads per run**	**Time per run**	**Cost per 1 million bases (in US$)**	**Advantages**	**Disadvantages**
Single-molecule real-time sequencing (Pacific Biosciences)	10,000 bp to 15,000 bp avg (14,000 bp N50); maximum read length >40,000 bases	87% single-read accuracy	50,000 per SMRT cell, or 500–1000 megabases	30 minutes to 4 hours	$0.13–$0.60	Longest read length. Fast. Detects 4mC, 5mC, 6mA.	Moderate throughput. Equipment can be very expensive.
Ion semiconductor (Ion Torrent sequencing)	up to 400 bp	98%	up to 80 million	2 hours	$1	Less expensive equipment. Fast.	Homopolymer errors.
Pyrosequencing (454)	700 bp	99.9%	1 million	24 hours	$10	Long read size. Fast.	Runs are expensive. Homopolymer errors.
Sequencing by synthesis (Illumina)	MiniSeq, NextSeq: 75-300 bp; MiSeq: 50-600 bp; HiSeq 2500: 50-500 bp; HiSeq 3/4000: 50-300 bp; HiSeq X: 300 bp	99.9% (Phred30)	MiniSeq/MiSeq: 1-25 Million; NextSeq: 130-00 Million, HiSeq 2500: 300 million - 2 billion, HiSeq 3/4000 2.5 billion, HiSeq X: 3 billion	1 to 11 days, depending upon sequencer and specified read length	$0.05 to $0.15	Potential for high sequence yield, depending upon sequencer model and desired application.	Equipment can be very expensive. Requires high concentrations of DNA.
Sequencing by ligation (SOLiD sequencing)	50+35 or 50+50 bp	99.9%	1.2 to 1.4 billion	1 to 2 weeks	$0.13	Low cost per base.	Slower than other methods. Has issues sequencing palindromic sequences.

Chain termination (Sanger sequencing)	400 to 900 bp	99.9%	N/A	20 minutes to 3 hours	$2400	Long individual reads. Useful for many applications.	More expensive and impractical for larger sequencing projects. This method also requires the time consuming step of plasmid cloning or PCR.

Massively Parallel Signature Sequencing (MPSS)

The first of the next-generation sequencing technologies, massively parallel signature sequencing (or MPSS), was developed in the 1990s at Lynx Therapeutics, a company founded in 1992 by Sydney Brenner and Sam Eletr. MPSS was a bead-based method that used a complex approach of adapter ligation followed by adapter decoding, reading the sequence in increments of four nucleotides. This method made it susceptible to sequence-specific bias or loss of specific sequences. Because the technology was so complex, MPSS was only performed 'in-house' by Lynx Therapeutics and no DNA sequencing machines were sold to independent laboratories. Lynx Therapeutics merged with Solexa (later acquired by Illumina) in 2004, leading to the development of sequencing-by-synthesis, a simpler approach acquired from Manteia Predictive Medicine, which rendered MPSS obsolete. However, the essential properties of the MPSS output were typical of later "next-generation" data types, including hundreds of thousands of short DNA sequences. In the case of MPSS, these were typically used for sequencing cDNA for measurements of gene expression levels.

Polony Sequencing

The Polony sequencing method, developed in the laboratory of George M. Church at Harvard, was among the first next-generation sequencing systems and was used to sequence a full *E. coli* genome in 2005. It combined an in vitro paired-tag library with emulsion PCR, an automated microscope, and ligation-based sequencing chemistry to sequence an *E. coli* genome at an accuracy of >99.9999% and a cost approximately 1/9 that of Sanger sequencing. The technology was licensed to Agencourt Biosciences, subsequently spun out into Agencourt Personal Genomics, and eventually incorporated into the Applied Biosystems SOLiD platform. Applied Biosystems was later acquired by Life Technologies, now part of Thermo Fisher Scientific.

454 Pyrosequencing

A parallelized version of pyrosequencing was developed by 454 Life Sciences, which has since been acquired by Roche Diagnostics. The method amplifies DNA inside water droplets in an oil solution (emulsion PCR), with each droplet containing a single DNA template attached to a single primer-coated bead that then forms a clonal colony. The sequencing machine contains many picoliter-volume wells each containing a single bead and sequencing enzymes. Pyrosequencing uses luciferase to generate light for detection of the individual nucleotides added to the nascent DNA, and the combined data are used to generate sequence read-outs. This technology provides intermediate read length and price per base compared to Sanger sequencing on one end and Solexa and SOLiD on the other.

Illumina (Solexa) Sequencing

Solexa, now part of Illumina, was founded by Shankar Balasubramanian and David Klenerman in 1998, and developed a sequencing method based on reversible dye-terminators technology, and engineered polymerases. The reversible terminated chemistry concept was invented by Bruno Canard and Simon Sarfati at the Pasteur Institute in Paris. It was developed internally at Solexa by those named on the relevant patents. In 2004, Solexa acquired the company Manteia Predictive Medicine in order to gain a massivelly parallel sequencing technology invented in 1997 by Pascal Mayer and Laurent Farinelli. It is based on "DNA Clusters" or "DNA colonies", which involves the clonal amplification of DNA on a surface. The cluster technology was co-acquired with Lynx Therapeutics of California. Solexa Ltd. later merged with Lynx to form Solexa Inc.

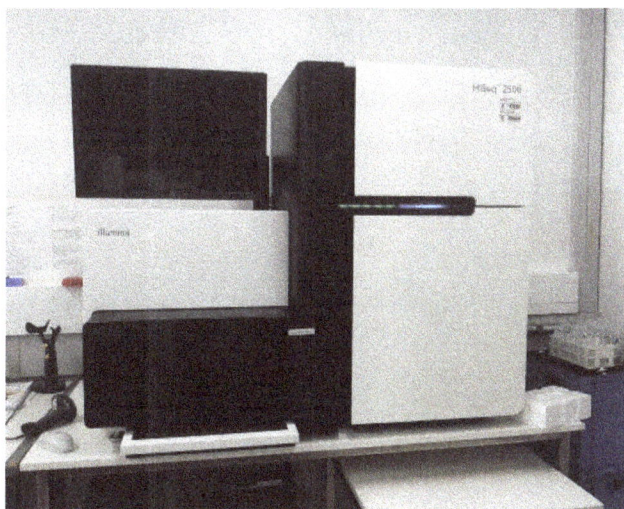

An Illumina HiSeq 2500 sequencer

In this method, DNA molecules and primers are first attached on a slide or flow cell and amplified with polymerase so that local clonal DNA colonies, later coined "DNA clusters", are formed. To determine the sequence, four types of reversible terminator bases (RT-bases) are added and non-incorporated nucleotides are washed away. A camera takes images of the fluorescently labeled nucleotides. Then the dye, along with the terminal 3' blocker, is chemically removed from the DNA, allowing for the next cycle to begin. Unlike pyrosequencing, the DNA chains are extended one nucleotide at a time and image acquisition can be performed at a delayed moment, allowing for very large arrays of DNA colonies to be captured by sequential images taken from a single camera.

Decoupling the enzymatic reaction and the image capture allows for optimal throughput and theoretically unlimited sequencing capacity. With an optimal configuration, the ultimately reachable instrument throughput is thus dictated solely by the analog-to-digital conversion rate of the camera, multiplied by the number of cameras and divided by the number of pixels per DNA colony required for visualizing them optimally (approximately 10 pixels/colony). In 2012, with cameras operating at more than 10 MHz A/D conversion rates and available optics, fluidics and enzymatics, throughput can be multiples of 1 million nucleotides/second, corresponding roughly to 1 human genome equivalent at 1x coverage per hour per instrument, and 1 human genome re-sequenced (at approx. 30x) per day per instrument (equipped with a single camera).

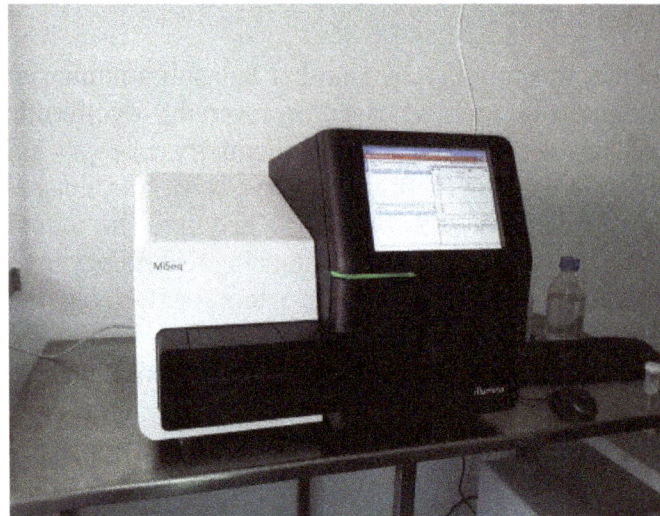

An Illumina MiSeq sequencer

SOLiD Sequencing

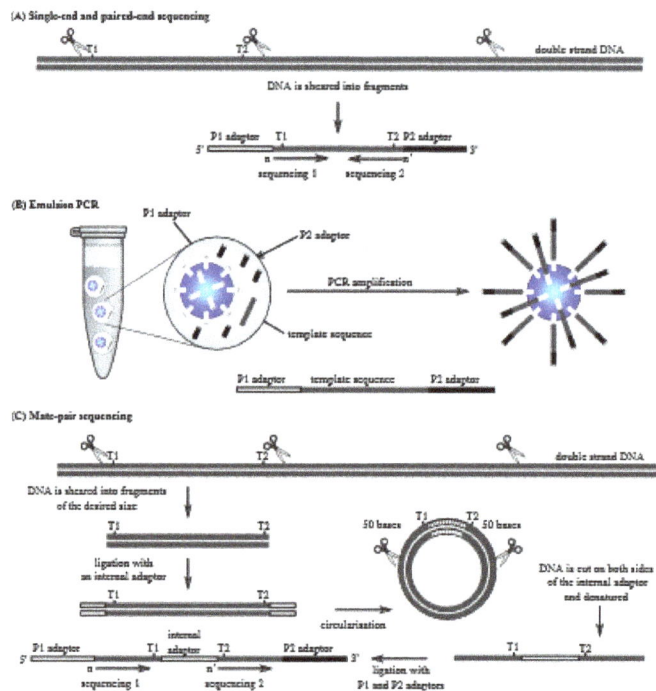

Library preparation for the SOLiD platform

Applied Biosystems' (now a Life Technologies brand) SOLiD technology employs sequencing by ligation. Here, a pool of all possible oligonucleotides of a fixed length are labeled according to the sequenced position. Oligonucleotides are annealed and ligated; the preferential ligation by DNA ligase for matching sequences results in a signal informative of the nucleotide at that position. Before sequencing, the DNA is amplified by emulsion PCR. The resulting beads, each containing single copies of the same DNA molecule, are deposited on a glass slide. The result is sequences of quantities and lengths comparable to Illumina sequencing. This sequencing by ligation method has been reported to have some issue sequencing palindromic sequences.

Ion Torrent Semiconductor Sequencing

Ion Torrent Systems Inc. (now owned by Life Technologies) developed a system based on using standard sequencing chemistry, but with a novel, semiconductor based detection system. This method of sequencing is based on the detection of hydrogen ions that are released during the polymerisation of DNA, as opposed to the optical methods used in other sequencing systems. A microwell containing a template DNA strand to be sequenced is flooded with a single type of nucleotide. If the introduced nucleotide is complementary to the leading template nucleotide it is incorporated into the growing complementary strand. This causes the release of a hydrogen ion that triggers a hypersensitive ion sensor, which indicates that a reaction has occurred. If homopolymer repeats are present in the template sequence multiple nucleotides will be incorporated in a single cycle. This leads to a corresponding number of released hydrogens and a proportionally higher electronic signal.

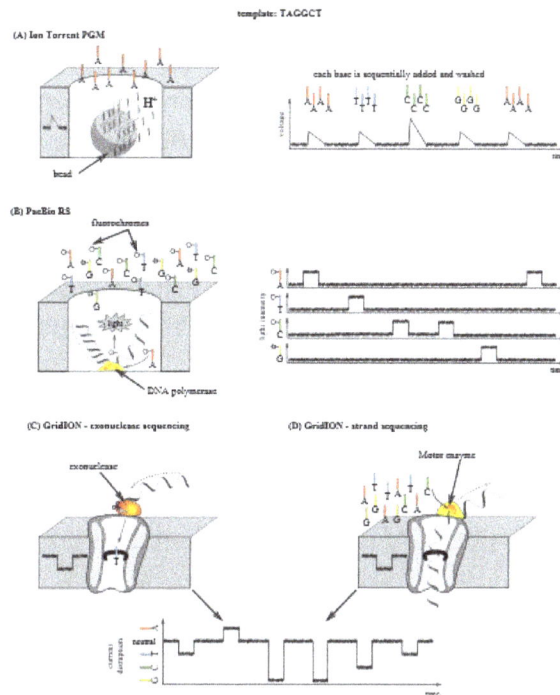

Sequencing of the TAGGCT template with IonTorrent, PacBioRS and GridION

DNA Nanoball Sequencing

DNA nanoball sequencing is a type of high throughput sequencing technology used to determine the entire genomic sequence of an organism. The company Complete Genomics uses this technology to sequence samples submitted by independent researchers. The method uses rolling circle replication to amplify small fragments of genomic DNA into DNA nanoballs. Unchained sequencing by ligation is then used to determine the nucleotide sequence. This method of DNA sequencing allows large numbers of DNA nanoballs to be sequenced per run and at low reagent costs compared to other next generation sequencing platforms. However, only short sequences of DNA are determined from each DNA nanoball which makes mapping the short reads to a reference genome difficult. This technology has been used for multiple genome sequencing projects and is scheduled to be used for more.

Heliscope Single Molecule Sequencing

Heliscope sequencing is a method of single-molecule sequencing developed by Helicos Biosciences. It uses DNA fragments with added poly-A tail adapters which are attached to the flow cell surface. The next steps involve extension-based sequencing with cyclic washes of the flow cell with fluorescently labeled nucleotides (one nucleotide type at a time, as with the Sanger method). The reads are performed by the Heliscope sequencer. The reads are short, averaging 35 bp. In 2009 a human genome was sequenced using the Heliscope, however in 2012 the company went bankrupt.

Single Molecule Real Time (SMRT) Sequencing

SMRT sequencing is based on the sequencing by synthesis approach. The DNA is synthesized in zero-mode wave-guides (ZMWs) – small well-like containers with the capturing tools located at the bottom of the well. The sequencing is performed with use of unmodified polymerase (attached to the ZMW bottom) and fluorescently labelled nucleotides flowing freely in the solution. The wells are constructed in a way that only the fluorescence occurring by the bottom of the well is detected. The fluorescent label is detached from the nucleotide upon its incorporation into the DNA strand, leaving an unmodified DNA strand. According to Pacific Biosciences (PacBio), the SMRT technology developer, this methodology allows detection of nucleotide modifications (such as cytosine methylation). This happens through the observation of polymerase kinetics. This approach allows reads of 20,000 nucleotides or more, with average read lengths of 5 kilobases. In 2015, Pacific Biosciences announced the launch of a new sequencing instrument called the Sequel System, with 1 million ZMWs compared to 150,000 ZMWs in the PacBio RS II instrument.

Nanopore DNA Sequencing

The DNA passing through the nanopore changes its ion current. This change is dependent on the shape, size and length of the DNA sequence. Each type of the nucleotide blocks the ion flow through the pore for a different period of time. The method does not require modified nucleotides and is performed in real time.

Early industrial research into this method was based on a technique called 'Exonuclease sequencing', where the readout of electrical signals occurring at nucleotides passing by alpha-hemolysin pores covalently bound with cyclodextrin. However the subsequently commercial method, 'strand sequencing' sequencing DNA bases in an intact strand. In 2014, Oxford Nanopore Technologies, a United Kingdom-based company, released the MinION portable DNA sequencer into an early access community and the first dataset became available in June 2014. The MinION was subsequently launched commercially in 2015 and the launch of the 'R9' nanopore in 2016 resulted in reports of performance improvements in accuracy and yield. in autumn 2016, the MinION handheld sequencer is capable of generating more than 2 Gigabases of sequencing data in one run. it is used in multiple scientific applications such as pathogen analysis/surveillance, metagenomics, cancer research and environmental analysis and a version has also been recently installed on the ISS by NASA.

Two main areas of nanopore sequencing in development are solid state nanopore sequencing, and protein based nanopore sequencing. Protein nanopore sequencing utilizes membrane protein complexes such as ∝-Hemolysin, MspA (Mycobacterium Smegmatis Porin A) or CssG, which show great promise given their ability to distinguish between individual and groups of nucleotides. In contrast, solid-state nanopore sequencing utilizes synthetic materials such as silicon nitride and aluminum oxide and it is preferred for its superior mechanical ability and thermal and chemical stability. The fabrication method is essential for this type of sequencing given that the nanopore array can contain hundreds of pores with diameters smaller than eight nanometers.

The concept originated from the idea that single stranded DNA or RNA molecules can be electrophoretically driven in a strict linear sequence through a biological pore that can be less than eight nanometers, and can be detected given that the molecules release an ionic current while moving through the pore. The pore contains a detection region capable of recognizing different bases, with each base generating various time specific signals corresponding to the sequence of bases as they cross the pore which are then evaluated. Precise control over the DNA transport through the pore is crucial for success. Various enzymes such as exonucleases and polymerases have been used to moderate this process by positioning them near the pore's entrance.

Methods in Development

DNA sequencing methods currently under development include reading the sequence as a DNA strand transits through nanopores (a method that is now commercial but subsequent generations such as solid-state nanopores are still in development), and microscopy-based techniques, such as atomic force microscopy or transmission electron microscopy that are used to identify the positions of individual nucleotides within long DNA fragments (>5,000 bp) by nucleotide labeling with heavier elements (e.g., halogens) for visual detection and recording. Third generation technologies aim to increase throughput and decrease the time to result and cost by eliminating the need for excessive reagents and harnessing the processivity of DNA polymerase.

Tunnelling Currents DNA Sequencing

Another approach uses measurements of the electrical tunnelling currents across single-strand DNA as it moves through a channel. Depending on its electronic structure, each base affects the tunnelling current differently, allowing differentiation between different bases.

The use of tunnelling currents has the potential to sequence orders of magnitude faster than ionic current methods and the sequencing of several DNA oligomers and micro-RNA has already been achieved.

Sequencing by Hybridization

Sequencing by hybridization is a non-enzymatic method that uses a DNA microarray. A single pool of DNA whose sequence is to be determined is fluorescently labeled and hybridized to an array containing known sequences. Strong hybridization signals from a given spot on the array identifies its sequence in the DNA being sequenced.

This method of sequencing utilizes binding characteristics of a library of short single stranded DNA molecules (oligonucleotides), also called DNA probes, to reconstruct a target DNA sequence. Non-specific hybrids are removed by washing and the target DNA is eluted. Hybrids are re-arranged such that the DNA sequence can be reconstructed. The benefit of this sequencing type is its ability to capture a large number of targets with a homogenous coverage. A large number of chemicals and starting DNA is usually required. However, with the advent of solution-based hybridization, much less equipment and chemicals are necessary.

Sequencing with Mass Spectrometry

Mass spectrometry may be used to determine DNA sequences. Matrix-assisted laser desorption ionization time-of-flight mass spectrometry, or MALDI-TOF MS, has specifically been investigated as an alternative method to gel electrophoresis for visualizing DNA fragments. With this method, DNA fragments generated by chain-termination sequencing reactions are compared by mass rather than by size. The mass of each nucleotide is different from the others and this difference is detectable by mass spectrometry. Single-nucleotide mutations in a fragment can be more easily detected with MS than by gel electrophoresis alone. MALDI-TOF MS can more easily detect differences between RNA fragments, so researchers may indirectly sequence DNA with MS-based methods by converting it to RNA first.

The higher resolution of DNA fragments permitted by MS-based methods is of special interest to researchers in forensic science, as they may wish to find single-nucleotide polymorphisms in human DNA samples to identify individuals. These samples may be highly degraded so forensic researchers often prefer mitochondrial DNA for its higher stability and applications for lineage studies. MS-based sequencing methods have been used to compare the sequences of human mitochondrial DNA from samples in a Federal Bureau of Investigation database and from bones found in mass graves of World War I soldiers.

Early chain-termination and TOF MS methods demonstrated read lengths of up to 100 base pairs. Researchers have been unable to exceed this average read size; like chain-termination sequencing alone, MS-based DNA sequencing may not be suitable for large *de novo* sequencing projects. Even so, a recent study did use the short sequence reads and mass spectroscopy to compare single-nucleotide polymorphisms in pathogenic *Streptococcus* strains.

Microfluidic Sanger Sequencing

In microfluidic Sanger sequencing the entire thermocycling amplification of DNA fragments as well as their separation by electrophoresis is done on a single glass wafer (approximately 10 cm in diameter) thus reducing the reagent usage as well as cost. In some instances researchers have shown that they can increase the throughput of conventional sequencing through the use of microchips. Research will still need to be done in order to make this use of technology effective.

Microscopy-based Techniques

This approach directly visualizes the sequence of DNA molecules using electron microscopy. The first identification of DNA base pairs within intact DNA molecules by enzymatically incorporating modified bases, which contain atoms of increased atomic number, direct visualization and identification of individually labeled bases within a synthetic 3,272 base-pair DNA molecule and a 7,249 base-pair viral genome has been demonstrated.

RNAP Sequencing

This method is based on use of RNA polymerase (RNAP), which is attached to a polystyrene bead. One end of DNA to be sequenced is attached to another bead, with both beads being placed in optical traps. RNAP motion during transcription brings the beads in closer and their relative distance changes, which can then be recorded at a single nucleotide resolution. The sequence is deduced based on the four readouts with lowered concentrations of each of the four nucleotide types, similarly to the Sanger method. A comparison is made between regions and sequence information is deduced by comparing the known sequence regions to the unknown sequence regions.

In Vitro Virus High-throughput Sequencing

A method has been developed to analyze full sets of protein interactions using a combination of 454 pyrosequencing and an *in vitro* virus mRNA display method. Specifically, this method covalently links proteins of interest to the mRNAs encoding them, then detects the mRNA pieces using reverse transcription PCRs. The mRNA may then be amplified and sequenced. The combined method was titled IVV-HiTSeq and can be performed under cell-free conditions, though its results may not be representative of *in vivo* conditions.

Sample Preparation

The success of a DNA sequencing protocol is dependent on the sample preparation. A successful DNA extraction will yield a sample with long, non-degraded strands of DNA which require further preparation according to the sequencing technology to be used. For Sanger sequencing, either cloning procedures or PCR are required prior to sequencing. In the case of next generation sequencing methods, library preparation is required before processing.

With the advent of next generation sequencing, Illumina and Roche 454 methods have become a common approach to transcriptomic studies (RNAseq). RNA can be extracted from tissues of interest and converted to complimentary DNA (cDNA) using reverse transcriptase—a DNA polymerase that synthesizes a complimentary DNA based on existing strands of RNA in a PCR-like manner. Complimentary DNA can be processed the same way as genomic DNA, allowing the expression levels of RNAs to be determined for the tissue selected.

Development Initiatives

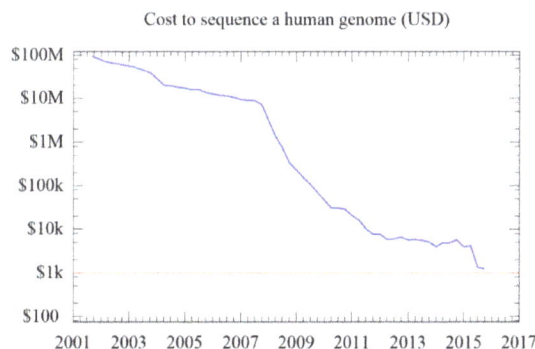

Total cost of sequencing a human genome over time as calculated by the NHGRI.

In October 2006, the X Prize Foundation established an initiative to promote the development of full genome sequencing technologies, called the Archon X Prize, intending to award $10 million to "the first Team that can build a device and use it to sequence 100 human genomes within 10 days or less, with an accuracy of no more than one error in every 100,000 bases sequenced, with sequences accurately covering at least 98% of the genome, and at a recurring cost of no more than $10,000 (US) per genome."

Each year the National Human Genome Research Institute, or NHGRI, promotes grants for new research and developments in genomics. 2010 grants and 2011 candidates include continuing work in microfluidic, polony and base-heavy sequencing methodologies.

Computational Challenges

The sequencing technologies described here produce raw data that needs to be assembled into longer sequences such as complete genomes (sequence assembly). There are many computational challenges to achieve this, such as the evaluation of the raw sequence data which is done by programs and algorithms such as Phred and Phrap. Other challenges have to deal with repetitive sequences that often prevent complete genome assemblies because they occur in many places of the genome. As a consequence, many sequences may not be assigned to particular chromosomes. The production of raw sequence data is only the beginning of its detailed bioinformatical analysis. Yet new methods for sequencing and correcting sequencing errors were developed.

Read Trimming

Sometimes, the raw reads produced by the sequencer are correct and precise only in a fraction of their length. Using the entire read may introduce artifacts in the downstream analyses like genome assembly, snp calling, or gene expression estimation. Two classes of trimming programs have been introduced, based on the window-based or the running-sum classes of algorithms. This is a partial list of the trimming algorithms currently available, specifying the algorithm class they belong to:

Read Trimming Algorithms		
Name of algorithm	**Type of algorithm**	**Link**
Cutadapt	Running sum	Cutadapt
ConDeTri	Window based	ConDeTri
ERNE-FILTER	Running sum	ERNE-FILTER
FASTX quality trimmer	Window based	FASTX quality trimmer
PRINSEQ	Window based	PRINSEQ
Trimmomatic	Window based	Trimmomatic
SolexaQA	Window based	SolexaQA
SolexaQA-BWA	Running sum	SolexaQA-BWA
Sickle	Window based	Sickle

Ethical Issues

Human genetics have been included within the field of bioethics since the early 1970s and the growth in the use of DNA sequencing (particularly high-throughput sequencing) has introduced a number of ethical issues. One key issue is the ownership of an individual's DNA and the data produced when that DNA is sequenced. Regarding the DNA molecule itself, the leading legal case on this topic, *Moore v. Regents of the University of California* (1990) ruled that individuals have no property rights to discarded cells or any profits made using these cells (for instance, as a patented cell line). However, individuals have a right to informed consent regarding removal and use of cells. Regarding the data produced through DNA sequencing, *Moore* gives the individual no rights to the information derived from their DNA.

As DNA sequencing becomes more widespread, the storage, security and sharing of genomic data has also become more important. For instance, one concern is that insurers may use an individual's genomic data to modify their quote, depending on the perceived future health of the individual based on their DNA. In May 2008, the Genetic Information Nondiscrimination Act (GINA) was signed in the United States, prohibiting discrimination on the basis of genetic information with respect to health insurance and employment. In 2012, the US Presidential Commission for the Study of Bioethical Issues reported that existing privacy legislation for DNA sequencing data such as GINA and the Health Insurance Portability and Accountability Act were insufficient, noting that whole-genome sequencing data was particularly sensitive, as it could be used to identify not only the individual from which the data was created, but also their relatives.

Ethical issues have also been raised by the increasing use of genetic variation screening, both in newborns, and in adults by companies such as 23andMe. It has been asserted that screening for genetic variations can be harmful, increasing anxiety in individuals who have been found to have an increased risk of disease. For example, in one case noted in *Time*, doctors screening an ill baby for genetic variants chose not to inform the parents of an unrelated variant linked to dementia due to the harm it would cause to the parents. However, a 2011 study in *The New England Journal of Medicine* has shown that individuals undergoing disease risk profiling did not show increased levels of anxiety.

Nucleic Acid Sequence

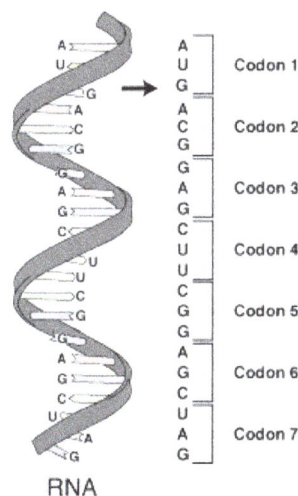

RNA

Ribonucleic acid

A series of codons in part of a mRNA molecule. Each codon consists of three nucleotides, usually representing a single amino acid.

A nucleic acid sequence is a succession of letters that indicate the order of nucleotides within a DNA (using GACT) or RNA (GACU) molecule. By convention, sequences are usually presented from the 5' end to the 3' end. For DNA, the sense strand is used. Because nucleic acids are normally linear (unbranched) polymers, specifying the sequence is equivalent to defining the covalent structure of the entire molecule. For this reason, the nucleic acid sequence is also termed the primary structure.

The sequence has capacity to represent information. Biological deoxyribonucleic acid represents the information which directs the functions of a living thing.

Nucleic acids also have a secondary structure and tertiary structure. Primary structure is sometimes mistakenly referred to as *primary sequence*. Conversely, there is no parallel concept of secondary or tertiary sequence.

Nucleotides

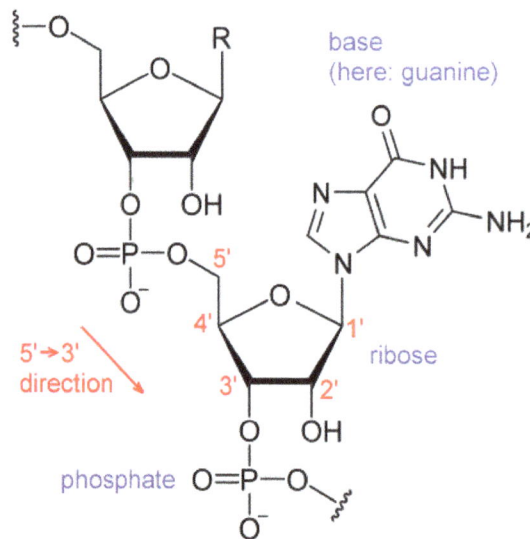

Chemical structure of RNA

Nucleic acids consist of a chain of linked units called nucleotides. Each nucleotide consists of three subunits: a phosphate group and a sugar (ribose in the case of RNA, deoxyribose in DNA) make up the backbone of the nucleic acid strand, and attached to the sugar is one of a set of nucleobases. The nucleobases are important in base pairing of strands to form higher-level secondary and tertiary structure such as the famed double helix.

The possible letters are *A*, *C*, *G*, and *T*, representing the four nucleotide bases of a DNA strand — adenine, cytosine, guanine, thymine — covalently linked to a phosphodiester backbone. In the typical case, the sequences are printed abutting one another without gaps, as in the sequence AAAGTCTGAC, read left to right in the 5' to 3' direction. With regards to transcription, a sequence is on the coding strand if it has the same order as the transcribed RNA.

One sequence can be complementary to another sequence, meaning that they have the base on each position in the complementary (i.e. A to T, C to G) and in the reverse order. For example, the complementary sequence to TTAC is GTAA. If one strand of the double-stranded DNA is considered the sense strand, then the other strand, considered the antisense strand, will have the complementary sequence to the sense strand.

Notation

While A, T, C, and G represent a particular nucleotide at a position, there are also letters that represent ambiguity which are used when more than one kind of nucleotide could occur at that position. The rules of the International Union of Pure and Applied Chemistry (IUPAC) are as follows:

- A = adenine
- C = cytosine
- G = guanine
- T = thymine
- R = G A (purine)
- Y = T C (pyrimidine)
- K = G T (keto)
- M = A C (amino)
- S = G C (strong bonds)
- W = A T (weak bonds)
- B = G T C (all but A)
- D = G A T (all but C)
- H = A C T (all but G)
- V = G C A (all but T)
- N = A G C T (any)

These symbols are also valid for RNA, except with U (uracil) replacing T (thymine).

Apart from adenine (A), cytosine (C), guanine (G), thymine (T) and uracil (U), DNA and RNA also contain bases that have been modified after the nucleic acid chain has been formed. In DNA, the most common modified base is 5-methylcytidine (m5C). In RNA, there are many modified bases, including pseudouridine (Ψ), dihydrouridine (D), inosine (I), ribothymidine (rT) and 7-methylguanosine (m7G). Hypoxanthine and xanthine are two of the many bases created through mutagen presence, both of them through deamination (replacement of the amine-group with a carbonyl-group). Hypoxanthine is produced from adenine, xanthine from guanine. Similarly, deamination of cytosine results in uracil.

Biological Significance

In biological systems, nucleic acids contain information which is used by a living cell to construct specific proteins. The sequence of nucleobases on a nucleic acid strand is translated by cell machinery into a sequence of amino acids making up a protein strand. Each group of three bases, called a codon, corresponds to a single amino acid, and there is a specific genetic code by which each possible combination of three bases corresponds to a specific amino acid.

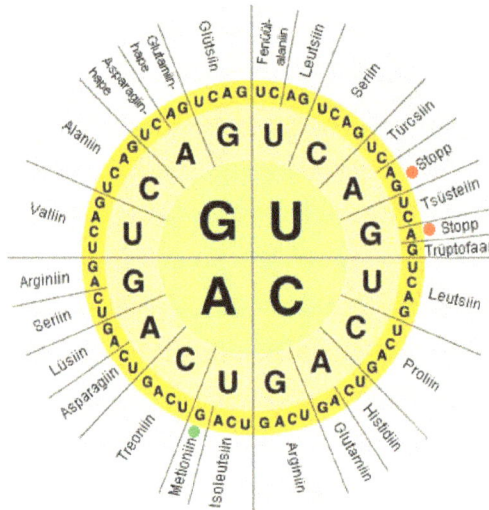

A depiction of the genetic code, by which the information contained in nucleic acids are translated into amino acid sequences in proteins.

The central dogma of molecular biology outlines the mechanism by which proteins are constructed using information contained in nucleic acids. DNA is transcribed into mRNA molecules, which travels to the ribosome where the mRNA is used as a template for the construction of the protein strand. Since nucleic acids can bind to molecules with complementary sequences, there is a distinction between "sense" sequences which code for proteins, and the complementary "antisense" sequence which is by itself nonfunctional, but can bind to the sense strand.

Sequence Determination

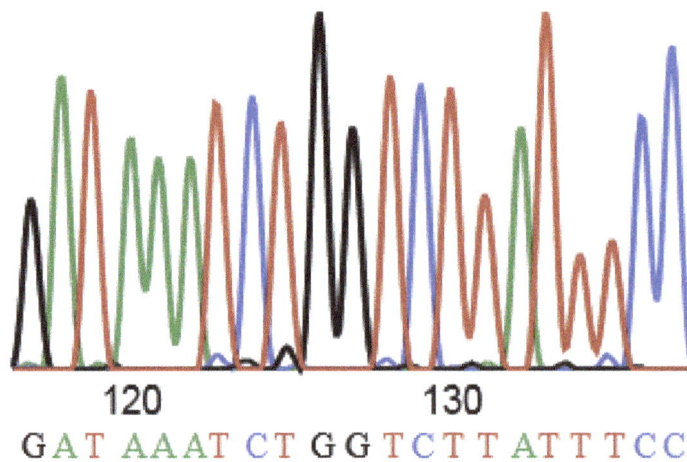

Electropherogram printout from automated sequencer for determining part of a DNA sequence

DNA sequencing is the process of determining the nucleotide sequence of a given DNA fragment. The sequence of the DNA of a living thing encodes the necessary information for that living thing to survive and reproduce. Therefore, determining the sequence is useful in fundamental research into why and how organisms live, as well as in applied subjects. Because of the importance of DNA to living things, knowledge of a DNA sequence may be useful in practically any biological research. For example, in medicine it can be used to identify, diagnose and potentially develop treatments for genetic diseases. Similarly, research into pathogens may lead to treatments for contagious diseases. Biotechnology is a burgeoning discipline, with the potential for many useful products and services.

RNA is not sequenced directly. Instead, it is copied to a DNA by reverse transcriptase, and this DNA is then sequenced.

Current sequencing methods rely on the discriminatory ability of DNA polymerases, and therefore can only distinguish four bases. An inosine (created from adenosine during RNA editing) is read as a G, and 5-methyl-cytosine (created from cytosine by DNA methylation) is read as a C. With current technology, it is difficult to sequence small amounts of DNA, as the signal is too weak to measure. This is overcome by polymerase chain reaction (PCR) amplification.

Digital Representation

Genetic sequence in digital format.

Once a nucleic acid sequence has been obtained from an organism, it is stored *in silico* in digital format. Digital genetic sequences may be stored in sequence databases, be analyzed be digitally altered and be used as templates for creating new actual DNA using artificial gene synthesis.

Sequence Analysis

Digital genetic sequences may be analyzed using the tools of bioinformatics to attempt to determine its function.

Genetic Testing

The DNA in an organism's genome can be analyzed to diagnose vulnerabilities to inherited diseases, and can also be used to determine a child's paternity (genetic father) or a person's ancestry. Normally, every person carries two variations of every gene, one inherited from their mother, the other inherited from their father. The human genome is believed to contain around 20,000 - 25,000 genes. In addition to studying chromosomes to the level of individual genes, genetic testing in a broader sense includes biochemical tests for the possible presence of genetic diseases, or mutant forms of genes associated with increased risk of developing genetic disorders.

Genetic testing identifies changes in chromosomes, genes, or proteins. Usually, testing is used to find changes that are associated with inherited disorders. The results of a genetic test can confirm or rule out a suspected genetic condition or help determine a person's chance of developing or passing on a genetic disorder. Several hundred genetic tests are currently in use, and more are being developed.

Sequence Alignment

In bioinformatics, a sequence alignment is a way of arranging the sequences of DNA, RNA, or protein to identify regions of similarity that may be due to functional, structural, or evolutionary relationships between the sequences. If two sequences in an alignment share a common ancestor, mismatches can be interpreted as point mutations and gaps as insertion or deletion mutations (indels) introduced in one or both lineages in the time since they diverged from one another. In sequence alignments of proteins, the degree of similarity between amino acids occupying a particular position in the sequence can be interpreted as a rough measure of how conserved a particular region or sequence motif is among lineages. The absence of substitutions, or the presence of only very conservative substitutions (that is, the substitution of amino acids whose side chains have similar biochemical properties) in a particular region of the sequence, suggest that this region has structural or functional importance. Although DNA and RNA nucleotide bases are more similar to each other than are amino acids, the conservation of base pairs can indicate a similar functional or structural role.

Computational phylogenetics makes extensive use of sequence alignments in the construction and interpretation of phylogenetic trees, which are used to classify the evolutionary relationships between homologous genes represented in the genomes of divergent species. The degree to which sequences in a query set differ is qualitatively related to the sequences' evolutionary distance from one another. Roughly speaking, high sequence identity suggests that the sequences in question have a comparatively young most recent common ancestor, while low identity suggests that the divergence is more ancient. This approximation, which reflects the "molecular clock" hypothesis that a roughly constant rate of evolutionary change can be used to extrapolate the elapsed time since two genes first diverged (that is, the coalescence time), assumes that the effects of mutation and selection are constant across sequence lineages. Therefore, it does not account for possible difference among organisms or species in the rates of DNA repair or the possible functional conservation of specific regions in a sequence. (In the case of nucleotide sequences, the molecular clock hypothesis in its most basic form also discounts the difference in acceptance rates between silent mutations that do not alter the meaning of a given codon and other mutations that result in a different amino acid being incorporated into the protein.) More statistically accurate methods allow the evolutionary rate on each branch of the phylogenetic tree to vary, thus producing better estimates of coalescence times for genes.

Sequence Motifs

Frequently the primary structure encodes motifs that are of functional importance. Some examples of sequence motifs are: the C/D and H/ACA boxes of snoRNAs, Sm binding site found in spliceosomal RNAs such as U1, U2, U4, U5, U6, U12 and U3, the Shine-Dalgarno sequence, the Kozak consensus sequence and the RNA polymerase III terminator.

Long Range Correlations

Peng found the existence of long-range correlations in the non-coding base pair sequences of DNA. In contrast, such correlations seem not to appear in coding DNA sequences.

Sequence Entropy

In Bioinformatics, a sequence entropy, also known as sequence complexity or information profile, is a numerical sequence providing a quantitative measure of the local complexity of a DNA sequence, independently of the direction of processing. The manipulations of the information profiles enable the analysis of the sequences using alignment-free techniques, such as for example in motif and rearrangements detection.

References

- Alberts B, Johnson A, Lewis J, Raff M, Roberts K, Walter P (2002). Molecular Biology of the Cell (Fourth ed.). New York: Garland Science. ISBN 978-0-8153-3218-3.

- Henig, Robin Marantz (2000). The Monk in the Garden: The Lost and Found Genius of Gregor Mendel, the Father of Genetics. Boston: Houghton Mifflin. pp. 1–9. ISBN 978-0395-97765-1.

- Judson, Horace (1979). The Eighth Day of Creation: Makers of the Revolution in Biology. Cold Spring Harbor Laboratory Press. pp. 51–169. ISBN 0-87969-477-7.

- Williams, George C. (2001). Adaptation and Natural Selection a Critique of Some Current Evolutionary Thought. ([Online-Ausg.]. ed.). Princeton: Princeton University Press. ISBN 9781400820108.

- Wagner EG; Altuvia S; Romby P (2002). "Antisense RNAs in bacteria and their genetic elements". Adv Genet. Advances in Genetics. 46: 361–98. doi:10.1016/S0065-2660(02)46013-0. ISBN 9780120176465. PMID 11931231.

- Harmon, Amy (24 February 2008). "Insurance Fears Lead Many to Shun DNA Tests". The New York Times. Retrieved 20 May 2015.

- "Privacy and Progress in Whole Genome Sequencing" (PDF). Presidential Commission for the Study of Bioethical Issues. Retrieved 20 May 2015.

- Hughes, Virginia. "It's Time To Stop Obsessing About the Dangers of Genetic Information". Slate Magazine. Retrieved 22 May 2015.

- "What kinds of gene mutations are possible?". Genetics Home Reference. United States National Library of Medicine. 11 May 2015. Retrieved 19 May 2015.

- Marlaire, Ruth (3 March 2015). "NASA Ames Reproduces the Building Blocks of Life in Laboratory". NASA. Retrieved 5 March 2015.

- Nuwer, Rachel (18 July 2015). "Counting All the DNA on Earth". The New York Times. New York: The New York Times Company. ISSN 0362-4331. Retrieved 2015-07-18.

Principles and Concepts of Genetics

The concepts of genetic variation, genetic drift, deletion, mutation and natural selection explain the physical and cognitive differences that exist in the human population at any given time. They also explain the reason that certain characteristics survive and others go extinct. This chapter discusses these key concepts in depth providing the reader with an awareness of the reasons behind why genetic research is one of the most rapidly growing fields of pure science.

Genetic Variation

Genetic variation is based on the variation in alleles of genes in a gene pool. It occurs both within and among populations, supported by individual carriers of the variant genes. Genetic variation is brought about by random mutation, which is a permanent change in the chemical structure of a gene.

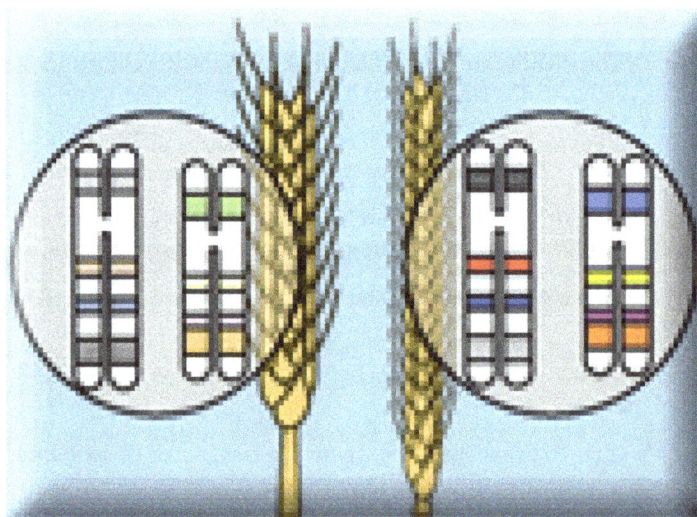

Genetic variation is a fact that a biological system – individual and population – is different over space. It is the base of the Genetic variability of different biological systems in space.

Among Individuals Within a Population

Genetic variation among individuals within a population can be identified at a variety of levels. It is possible to identify genetic variation from observations of phenotypic variation in either quantitative traits (traits that vary continuously and are coded for by many genes (e.g., leg length in dogs)) or discrete traits (traits that fall into discrete categories and are coded for by one or a few genes (e.g., white, pink, red petal color in certain flowers)).

Genetic variation can also be identified by examining variation at the level of enzymes using the process of protein electrophoresis. Polymorphic genes have more than one allele at each locus. Half of the genes that code for enzymes in insects and plants may be polymorphic, whereas polymorphisms are less common in vertebrates.

Ultimately, genetic variation is caused by variation in the order of bases in the nucleotides in genes. New technology now allows scientists to directly sequence DNA which has identified even more genetic variation than was previously detected by protein electrophoresis. Examination of DNA has shown genetic variation in both coding regions and in the non-coding intron region of genes.

Genetic variation will result in phenotypic variation if variation in the order of nucleotides in the DNA sequence results in a difference in the order of amino acids in proteins coded by that DNA sequence, and if the resultant differences in amino acid sequence influence the shape, and thus the function of the enzyme.

Between Populations

Geographic variation in genes often occurs among populations living in different locations. Geographic variation may be due to differences in selective pressures or to genetic drift.

Measurement

Genetic variation within a population is commonly measured as the percentage of gene loci that are polymorphic or the percentage of gene loci in individuals that are heterozygous.

Sources

A range of variability in the mussel Donax variabilis

Random mutations are the ultimate source of genetic variation. Mutations are likely to be rare and most mutations are neutral or deleterious, but in some instances the new alleles can be favored by natural selection.

Polyploidy is an example of chromosomal mutation. Polyploidy is a condition wherein organisms have three or more sets of genetic variation (3n or more). The mutation is started off by a parent, as the parent mates the offspring now has a chance to receive that mutation trait also. Now when that mutated offspring is ready to mate they now have the chance of passing on that trait to their offspring. This process begins the first generation of mutated offspring.

Variability of walnuts

Crossing over and random segregation during meiosis can result in the production of new alleles or new combinations of alleles. Furthermore, random fertilization also contributes to variation.

Variation and recombination can be facilitated by transposable and transposed genetic elements, commonly known as endogenous retroviruses, LINEs, SINEs, etc.

For a given genome of a multicellular organism, genetic variation may be acquired in somatic cells or inherited through the germline.

Forms

Genetic variation can be divided into different forms according to the size and type of genomic variation underpinning genetic change. These include:

- Small-scale sequence variation (< 1Kbp):

    o Substitution

    o Indels

- Large-scale Structural variation (>1Kbp):

    o Copy number variation:

        ☐ Copy number loss

        ☐ Copy number gain

    o Rearrangement:

□ Translocation

□ Inversion

□ Segmental acquired uniparental disomy

- Numerical variation (whole chromosomes or genomes):

  o Polyploidy

  o Aneuploidy

Maintenance in Populations

A variety of factors maintain genetic variation in populations. Potentially harmful recessive alleles can be hidden from selection in the heterozygous individuals in populations of diploid organisms (recessive alleles are only expressed in the less common homozygous individuals). Natural selection can also maintain genetic variation in balanced polymorphisms. Balanced polymorphisms may occur when heterozygotes are favored or when selection is frequency dependent.

Genetic Drift

Biologist and statistician Ronald Fisher

Genetic drift (also known as allelic drift or the Sewall Wright effect after biologist Sewall Wright) is the change in the frequency of a gene variant (allele) in a population due to random sampling of organisms. The alleles in the offspring are a sample of those in the parents, and chance has a role in determining whether a given individual survives and reproduces. A population's allele frequency is the fraction of the copies of one gene that share a particular form. Genetic drift may cause gene variants to disappear completely and thereby reduce genetic variation.

When there are few copies of an allele, the effect of genetic drift is larger, and when there are many copies the effect is smaller. In the early 20th century, vigorous debates occurred over the relative importance of natural selection versus neutral processes, including genetic drift. Ronald Fisher, who explained natural selection using Mendelian genetics, held the view that genetic drift plays at the most a minor role in evolution, and this remained the dominant view for several decades. In 1968, population geneticist Motoo Kimura rekindled the debate with his neutral theory of molecular evolution, which claims that most instances where a genetic change spreads across a population (although not necessarily changes in phenotypes) are caused by genetic drift acting on neutral mutations.

Analogy with Marbles in a Jar

The process of genetic drift can be illustrated using 20 marbles in a jar to represent 20 organisms in a population. Consider this jar of marbles as the starting population. Half of the marbles in the jar are red and half blue, and both colours correspond to two different alleles of one gene in the population. In each new generation the organisms reproduce at random. To represent this reproduction, randomly select a marble from the original jar and deposit a new marble with the same colour as its "offspring" into a new jar. (The selected marble remains in the original jar.) Repeat this process until there are 20 new marbles in the second jar. The second jar then contains a second generation of "offspring," consisting of 20 marbles of various colours. Unless the second jar contains exactly 10 red marbles and 10 blue marbles, a random shift occurred in the allele frequencies.

Repeat this process a number of times, randomly reproducing each generation of marbles to form the next. The numbers of red and blue marbles picked each generation fluctuates; sometimes more red and sometimes more blue. This fluctuation is analogous to genetic drift – a change in the population's allele frequency resulting from a random variation in the distribution of alleles from one generation to the next.

It is even possible that in any one generation no marbles of a particular colour are chosen, meaning they have no offspring. In this example, if no red marbles are selected, the jar representing the new generation contains only blue offspring. If this happens, the red allele has been lost permanently in the population, while the remaining blue allele has become fixed: all future generations are entirely blue. In small populations, fixation can occur in just a few generations.

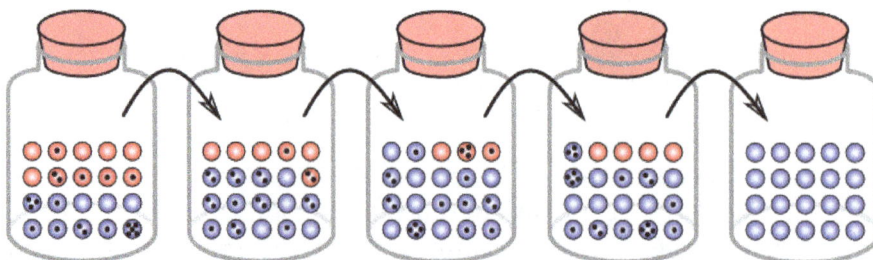

In this simulation, there is fixation in the blue "allele" within five generations.

Probability and Allele Frequency

The mechanisms of genetic drift can be illustrated with a simplified example. Consider a very large colony of bacteria isolated in a drop of solution. The bacteria are genetically identical except for a single gene with two alleles labeled A and B. A and B are neutral alleles meaning that they do not affect the bacteria's ability to survive and reproduce; all bacteria in this colony are equally likely to survive and reproduce. Suppose that half the bacteria have allele A and the other half have allele B. Thus A and B each have allele frequency 1/2.

The drop of solution then shrinks until it has only enough food to sustain four bacteria. All other bacteria die without reproducing. Among the four who survive, there are sixteen possible combinations for the A and B alleles:

$$(A\text{-}A\text{-}A\text{-}A), (B\text{-}A\text{-}A\text{-}A), (A\text{-}B\text{-}A\text{-}A), (B\text{-}B\text{-}A\text{-}A),$$
$$(A\text{-}A\text{-}B\text{-}A), (B\text{-}A\text{-}B\text{-}A), (A\text{-}B\text{-}B\text{-}A), (B\text{-}B\text{-}B\text{-}A),$$
$$(A\text{-}A\text{-}A\text{-}B), (B\text{-}A\text{-}A\text{-}B), (A\text{-}B\text{-}A\text{-}B), (B\text{-}B\text{-}A\text{-}B),$$
$$(A\text{-}A\text{-}B\text{-}B), (B\text{-}A\text{-}B\text{-}B), (A\text{-}B\text{-}B\text{-}B), (B\text{-}B\text{-}B\text{-}B).$$

Since all bacteria in the original solution are equally likely to survive when the solution shrinks, the four survivors are a random sample from the original colony. The probability that each of the four survivors has a given allele is 1/2, and so the probability that any particular allele combination occurs when the solution shrinks is

$$\frac{1}{2} \cdot \frac{1}{2} \cdot \frac{1}{2} \cdot \frac{1}{2} = \frac{1}{16}.$$

(The original population size is so large that the sampling effectively happens without replacement). In other words, each of the sixteen possible allele combinations is equally likely to occur, with probability 1/16.

Counting the combinations with the same number of A and B, we get the following table.

A	B	Combinations	Probability
4	0	1	1/16
3	1	4	4/16
2	2	6	6/16
1	3	4	4/16
0	4	1	1/16

As shown in the table, the total number of possible combinations to have an equal (conserved) number of A and B alleles is six, and its probability is 6/16. The total number of possible alternative combinations is ten, and the probability of unequal number of A and B alleles is 10/16. Thus, although the original colony began with an equal number of A and B alleles, chances are that the number of alleles in the remaining population of four members will not be equal. In the latter case, genetic drift has occurred because the population's allele frequencies have changed due to random sampling. In this example the population contracted to just four random survivors, a phenomenon known as population bottleneck.

The probabilities for the number of copies of allele A (or B) that survive (given in the last column of the above table) can be calculated directly from the binomial distribution where the "success" probability (probability of a given allele being present) is $1/2$ (i.e., the probability that there are k copies of A (or B) alleles in the combination) is given by

$$\binom{n}{k}\left(\frac{1}{2}\right)^k\left(1-\frac{1}{2}\right)^{n-k} = \binom{n}{k}\left(\frac{1}{2}\right)^n$$

where $n=4$ is the number of surviving bacteria.

Mathematical Models of Genetic Drift

Mathematical models of genetic drift can be designed using either branching processes or a diffusion equation describing changes in allele frequency in an idealised population.

Wright–fisher Model

Think of a gene with two alleles, A or B. In diploid populations consisting of N individuals there are $2N$ copies of each gene. An individual can have two copies of the same allele or two different alleles. We can call the frequency of one allele p and the frequency of the other q. The Wright–Fisher model (named after Sewall Wright and Ronald Fisher) assumes that generations do not overlap (for example, annual plants have exactly one generation per year) and that each copy of the gene found in the new generation is drawn independently at random from all copies of the gene in the old generation. The formula to calculate the probability of obtaining k copies of an allele that had frequency p in the last generation is then

$$\frac{(2N)!}{k!(2N-k)!}p^k q^{2N-k}$$

where the symbol "!" signifies the factorial function. This expression can also be formulated using the binomial coefficient,

$$\binom{2N}{k}p^k q^{2N-k}$$

Moran Model

The Moran model assumes overlapping generations. At each time step, one individual is chosen to reproduce and one individual is chosen to die. So in each timestep, the number of copies of a given allele can go up by one, go down by one, or can stay the same. This means that the transition matrix is tridiagonal, which means that mathematical solutions are easier for the Moran model than for the Wright–Fisher model. On the other hand, computer simulations are usually easier to perform using the Wright–Fisher model, because fewer time steps need to be calculated. In the Moran model, it takes N timesteps to get through one generation, where N is the effective population size. In the Wright–Fisher model, it takes just one.

In practice, the Moran model and Wright–Fisher model give qualitatively similar results, but genetic drift runs twice as fast in the Moran model.

Other Models of Drift

If the variance in the number of offspring is much greater than that given by the binomial distribution assumed by the Wright–Fisher model, then given the same overall speed of genetic drift (the variance effective population size), genetic drift is a less powerful force compared to selection. Even for the same variance, if higher moments of the offspring number distribution exceed those of the binomial distribution then again the force of genetic drift is substantially weakened.

Random Effects Other than Sampling Error

Random changes in allele frequencies can also be caused by effects other than sampling error, for example random changes in selection pressure.

One important alternative source of stochasticity, perhaps more important than genetic drift, is genetic draft. Genetic draft is the effect on a locus by selection on linked loci. The mathematical properties of genetic draft are different from those of genetic drift. The direction of the random change in allele frequency is autocorrelated across generations.

Drift and Fixation

The Hardy–Weinberg principle states that within sufficiently large populations, the allele frequencies remain constant from one generation to the next unless the equilibrium is disturbed by migration, genetic mutations, or selection.

However, in finite populations, no new alleles are gained from the random sampling of alleles passed to the next generation, but the sampling can cause an existing allele to disappear. Because random sampling can remove, but not replace, an allele, and because random declines or increases in allele frequency influence expected allele distributions for the next generation, genetic drift drives a population towards genetic uniformity over time. When an allele reaches a frequency of 1 (100%) it is said to be "fixed" in the population and when an allele reaches a frequency of 0 (0%) it is lost. Smaller populations achieve fixation faster, whereas in the limit of an infinite population, fixation is not achieved. Once an allele becomes fixed, genetic drift comes to a halt, and the allele frequency cannot change unless a new allele is introduced in the population via mutation or gene flow. Thus even while genetic drift is a random, directionless process, it acts to eliminate genetic variation over time.

Rate of Allele Frequency Change Due to Drift

Assuming genetic drift is the only evolutionary force acting on an allele, after t generations in many replicated populations, starting with allele frequencies of p and q, the variance in allele frequency across those populations is

$$V_t \approx pq\left(1-\exp\left(-\frac{t}{2N_e}\right)\right)$$

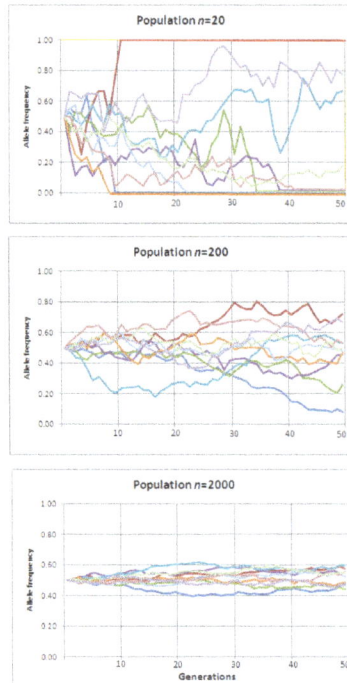

Ten simulations of random genetic drift of a single given allele with an initial frequency distribution 0.5 measured over the course of 50 generations, repeated in three reproductively synchronous populations of different sizes. In these simulations, alleles drift to loss or fixation (frequency of 0.0 or 1.0) only in the smallest population.

Time to Fixation or Loss

Assuming genetic drift is the only evolutionary force acting on an allele, at any given time the probability that an allele will eventually become fixed in the population is simply its frequency in the population at that time. For example, if the frequency p for allele A is 75% and the frequency q for allele B is 25%, then given unlimited time the probability A will ultimately become fixed in the population is 75% and the probability that B will become fixed is 25%.

The expected number of generations for fixation to occur is proportional to the population size, such that fixation is predicted to occur much more rapidly in smaller populations. Normally the effective population size, which is smaller than the total population, is used to determine these probabilities. The effective population (N_e) takes into account factors such as the level of inbreeding, the stage of the lifecycle in which the population is the smallest, and the fact that some neutral genes are genetically linked to others that are under selection. The effective population size may not be the same for every gene in the same population.

One forward-looking formula used for approximating the expected time before a neutral allele becomes fixed through genetic drift, according to the Wright–Fisher model, is

$$\overline{T}_{\text{fixed}} = \frac{-4N_e(1-p)\ln(1-p)}{p}$$

where T is the number of generations, N_e is the effective population size, and p is the initial frequency for the given allele. The result is the number of generations expected to pass before fixation occurs for a given allele in a population with given size (N_e) and allele frequency (p).

The expected time for the neutral allele to be lost through genetic drift can be calculated as

$$\overline{T}_{lost} = \frac{-4N_e p}{1-p} \ln p.$$

When a mutation appears only once in a population large enough for the initial frequency to be negligible, the formulas can be simplified to

$$\overline{T}_{fixed} = 4N_e$$

for average number of generations expected before fixation of a neutral mutation, and

$$\overline{T}_{lost} = 2\left(\frac{N_e}{N}\right)\ln(2N)$$

for the average number of generations expected before the loss of a neutral mutation.

Time to Loss with Both Drift and Mutation

The formulae above apply to an allele that is already present in a population, and which is subject to neither mutation nor natural selection. If an allele is lost by mutation much more often than it is gained by mutation, then mutation, as well as drift, may influence the time to loss. If the allele prone to mutational loss begins as fixed in the population, and is lost by mutation at rate m per replication, then the expected time in generations until its loss in a haploid population is given by

$$\overline{T}_{lost} \approx \begin{cases} \dfrac{1}{m}, & \text{if } mN_e \ll 1 \\[2ex] \dfrac{\ln(mN_e)+\gamma}{m} & \text{if } mN_e \gg 1 \end{cases}$$

where γ is equal to Euler's constant. The first approximation represents the waiting time until the first mutant destined for loss, with loss then occurring relatively rapidly by genetic drift, taking time $N_e \ll 1/m$. The second approximation represents the time needed for deterministic loss by mutation accumulation. In both cases, the time to fixation is dominated by mutation via the term $1/m$, and is less affected by the effective population size.

Genetic Drift Versus Natural Selection

In natural populations, genetic drift and natural selection do not act in isolation; both forces are always at play, together with mutation and migration. Neutral evolution is the product of both mutation and drift, not of drift alone. Similarly, even when selection overwhelms genetic drift, it can only act on variation that mutation provides.

While natural selection has a direction, guiding evolution towards heritable adaptations to the current environment, genetic drift has no direction and is guided only by the mathematics of chance.

As a result, drift acts upon the genotypic frequencies within a population without regard to their phenotypic effects. In contrast, selection favors the spread of alleles whose phenotypic effects increase survival and/or reproduction of their carriers, lowers the frequencies of alleles that cause unfavorable traits, and ignores those that are neutral.

The law of large numbers predicts that when the absolute number of copies of the allele is small (e.g., in small populations), the magnitude of drift on allele frequencies per generation is larger. The magnitude of drift is large enough to overwhelm selection at any allele frequency when the selection coefficient is less than 1 divided by the effective population size. Non-adaptive evolution resulting from the product of mutation and genetic drift is therefore considered to be a consequential mechanism of evolutionary change primarily within small, isolated populations. The mathematics of genetic drift depend on the effective population size, but it is not clear how this is related to the actual number of individuals in a population. Genetic linkage to other genes that are under selection can reduce the effective population size experienced by a neutral allele. With a higher recombination rate, linkage decreases and with it this local effect on effective population size. This effect is visible in molecular data as a correlation between local recombination rate and genetic diversity, and negative correlation between gene density and diversity at noncoding DNA regions. Stochasticity associated with linkage to other genes that are under selection is not the same as sampling error, and is sometimes known as genetic draft in order to distinguish it from genetic drift.

When the allele frequency is very small, drift can also overpower selection even in large populations. For example, while disadvantageous mutations are usually eliminated quickly in large populations, new advantageous mutations are almost as vulnerable to loss through genetic drift as are neutral mutations. Not until the allele frequency for the advantageous mutation reaches a certain threshold will genetic drift have no effect.

Population Bottleneck

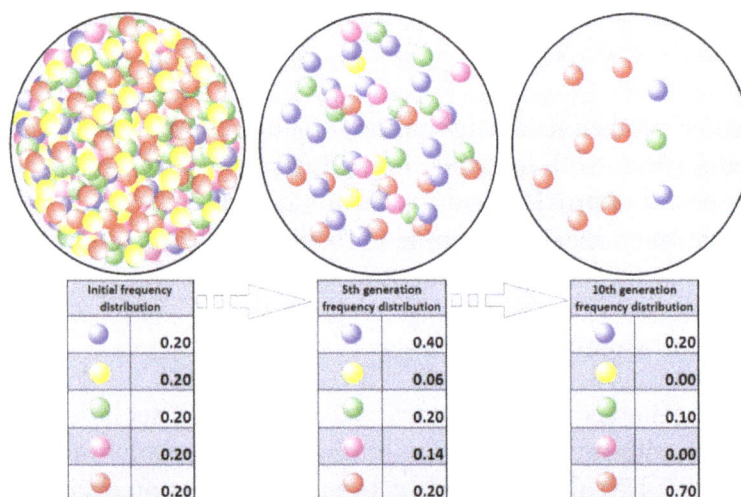

Initial frequency distribution		5th generation frequency distribution		10th generation frequency distribution	
🔵	0.20	🔵	0.40	🔵	0.20
🟡	0.20	🟡	0.06	🟡	0.00
🟢	0.20	🟢	0.20	🟢	0.10
🔴	0.20	🔴	0.14	🔴	0.00
🟠	0.20	🟠	0.20	🟠	0.70

Changes in a population's allele frequency following a population bottleneck: the rapid and radical decline in population size has reduced the population's genetic variation.

A population bottleneck is when a population contracts to a significantly smaller size over a short period of time due to some random environmental event. In a true population bottleneck, the odds for survival of any member of the population are purely random, and are not improved by any particular inherent genetic advantage. The bottleneck can result in radical changes in allele frequencies, completely independent of selection.

The impact of a population bottleneck can be sustained, even when the bottleneck is caused by a one-time event such as a natural catastrophe. An interesting example of a bottleneck causing unusual genetic distribution is the relatively high proportion of individuals with total rod cell color blindness (achromatopsia) on Pingelap atoll in Micronesia. After a bottleneck, inbreeding increases. This increases the damage done by recessive deleterious mutations, in a process known as inbreeding depression. The worst of these mutations are selected against, leading to the loss of other alleles that are genetically linked to them, in a process of background selection. For recessive harmful mutations, this selection can be enhanced as a consequence of the bottleneck, due to genetic purging. This leads to a further loss of genetic diversity. In addition, a sustained reduction in population size increases the likelihood of further allele fluctuations from drift in generations to come.

A population's genetic variation can be greatly reduced by a bottleneck, and even beneficial adaptations may be permanently eliminated. The loss of variation leaves the surviving population vulnerable to any new selection pressures such as disease, climate change or shift in the available food source, because adapting in response to environmental changes requires sufficient genetic variation in the population for natural selection to take place.

There have been many known cases of population bottleneck in the recent past. Prior to the arrival of Europeans, North American prairies were habitat for millions of greater prairie chickens. In Illinois alone, their numbers plummeted from about 100 million birds in 1900 to about 50 birds in the 1990s. The declines in population resulted from hunting and habitat destruction, but the random consequence has been a loss of most of the species' genetic diversity. DNA analysis comparing birds from the mid century to birds in the 1990s documents a steep decline in the genetic variation in just in the latter few decades. Currently the greater prairie chicken is experiencing low reproductive success. However, bottleneck and genetic drift can lead to a genetic loss that increases fitness as seen in Ehrlichia.

Over-hunting also caused a severe population bottleneck in the northern elephant seal in the 19th century. Their resulting decline in genetic variation can be deduced by comparing it to that of the southern elephant seal, which were not so aggressively hunted.

Founder Effect

The founder effect is a special case of a population bottleneck, occurring when a small group in a population splinters off from the original population and forms a new one. The random sample of alleles in the just formed new colony is expected to grossly misrepresent the original population in at least some respects. It is even possible that the number of alleles for some genes in the original population is larger than the number of gene copies in the founders, making complete representation impossible. When a newly formed colony is small, its founders can strongly affect the population's genetic make-up far into the future.

A well-documented example is found in the Amish migration to Pennsylvania in 1744. Two members of the new colony shared the recessive allele for Ellis–van Creveld syndrome. Members of the colony and their descendants tend to be religious isolates and remain relatively insular. As a result of many generations of inbreeding, Ellis-van Creveld syndrome is now much more prevalent among the Amish than in the general population.

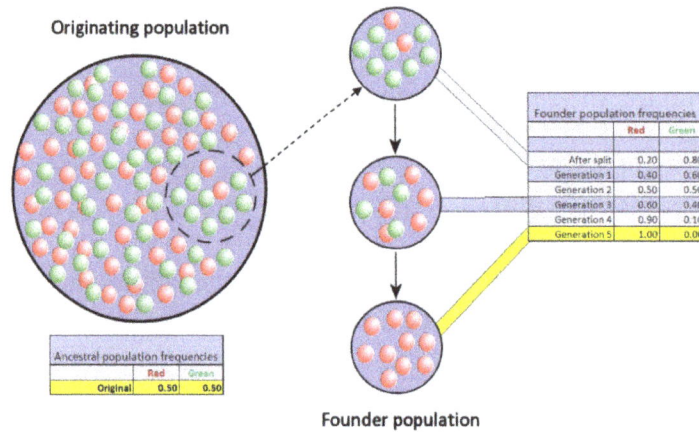

When very few members of a population migrate to form a separate new population, the founder effect occurs. For a period after the foundation, the small population experiences intensive drift. In the figure this results in fixation of the red allele.

The difference in gene frequencies between the original population and colony may also trigger the two groups to diverge significantly over the course of many generations. As the difference, or genetic distance, increases, the two separated populations may become distinct, both genetically and phenetically, although not only genetic drift but also natural selection, gene flow, and mutation contribute to this divergence. This potential for relatively rapid changes in the colony's gene frequency led most scientists to consider the founder effect (and by extension, genetic drift) a significant driving force in the evolution of new species. Sewall Wright was the first to attach this significance to random drift and small, newly isolated populations with his shifting balance theory of speciation. Following after Wright, Ernst Mayr created many persuasive models to show that the decline in genetic variation and small population size following the founder effect were critically important for new species to develop. However, there is much less support for this view today since the hypothesis has been tested repeatedly through experimental research and the results have been equivocal at best.

Founder effect was first well-investigated in the USSR by Soviet scientists Lisovskiy V.V., Kuznetsov M.A. and Nikolay Dubinin.

History of the Concept

The concept of genetic drift was first introduced by one of the founders in the field of population genetics, Sewall Wright. His first use of the term "drift" was in 1929, though at the time he was using it in the sense of a directed process of change, or natural selection. Random drift by means of sampling error came to be known as the "Sewall–Wright effect," though he was never entirely comfortable to see his name given to it. Wright referred to all changes in allele frequency as either "steady drift" (e.g., selection) or "random drift" (e.g., sampling error). "Drift" came to be adopted as a technical term in the stochastic sense exclusively. Today it is usually defined still more narrowly, in terms of sampling error, although this narrow definition is not universal. Wright wrote that the "restriction of "random drift" or even "drift" to only one component, the effects of accidents of sampling, tends to lead to confusion." Sewall Wright considered the process of random genetic drift by means of sampling error equivalent to that by means of inbreeding, but later work has shown them to be distinct.

In the early days of the modern evolutionary synthesis, scientists were just beginning to blend the new science of population genetics with Charles Darwin's theory of natural selection. Working within this new framework, Wright focused on the effects of inbreeding on small relatively isolated populations. He introduced the concept of an adaptive landscape in which phenomena such as cross breeding and genetic drift in small populations could push them away from adaptive peaks, which in turn allow natural selection to push them towards new adaptive peaks. Wright thought smaller populations were more suited for natural selection because "inbreeding was sufficiently intense to create new interaction systems through random drift but not intense enough to cause random nonadaptive fixation of genes."

Wright's views on the role of genetic drift in the evolutionary scheme were controversial almost from the very beginning. One of the most vociferous and influential critics was colleague Ronald Fisher. Fisher conceded genetic drift played some role in evolution, but an insignificant one. Fisher has been accused of misunderstanding Wright's views because in his criticisms Fisher seemed to argue Wright had rejected selection almost entirely. To Fisher, viewing the process of evolution as a long, steady, adaptive progression was the only way to explain the ever-increasing complexity from simpler forms. But the debates have continued between the "gradualists" and those who lean more toward the Wright model of evolution where selection and drift together play an important role.

In 1968, Motoo Kimura rekindled the debate with his neutral theory of molecular evolution, which claims that most of the genetic changes are caused by genetic drift acting on neutral mutations.

The role of genetic drift by means of sampling error in evolution has been criticized by John H. Gillespie and William B. Provine, who argue that selection on linked sites is a more important stochastic force.

Deletion (Genetics)

In genetics, a deletion (also called gene deletion, deficiency, or deletion mutation) (sign: Δ) is a mutation (a genetic aberration) in which a part of a chromosome or a sequence of DNA is lost during DNA replication. Any number of nucleotides can be deleted, from a single base to an entire piece of chromosome. The smallest single base deletion mutations are believed occur by a single base flipping in the template DNA, followed by template DNA strand slippage, within the DNA polymerase active site. Deletions can be caused by errors in chromosomal crossover during meiosis, which causes several serious genetic diseases. Deletions that do not occur in multiples of three bases can cause a frameshift by changing the 3-nucleotide protein reading frame of the genetic sequence.

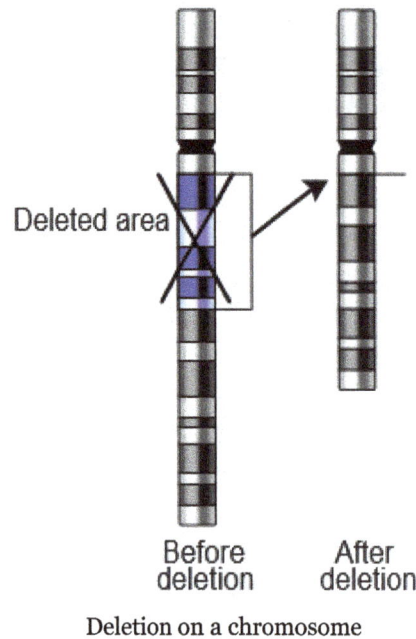

Deletion on a chromosome

Causes

Causes include the following:

- Losses from translocation

- Chromosomal crossovers within a chromosomal inversion

- Unequal crossing over

- Breaking without rejoining

For synapsis to occur between a chromosome with a large intercalary deficiency and a normal complete homolog, the unpaired region of the normal homolog must loop out of the linear structure into a deletion or compensation loop.

Types

Types of deletion include the following:

- *Terminal Deletion'* — a deletion that occurs towards the end of a chromosome.

- Intercalary Deletion / Interstitial Addition — a deletion that occurs from the interior of a chromosome.

- Microdeletion — a relatively small amount of deletion (up to 5Mb that could include a dozen genes).

Microdeletion is usually found in children with physical abnormalities. A large amount of deletion would result in immediate abortion (miscarriage).

Effects

Small deletions are less likely to be fatal; large deletions are usually fatal — there are always variations based on which genes are lost. Some medium-sized deletions lead to recognizable human disorders, e.g. Williams syndrome.

Deletion of a number of pairs that is not evenly divisible by three will lead to a frameshift mutation, causing all of the codons occurring after the deletion to be read incorrectly during translation, producing a severely altered and potentially nonfunctional protein. In contrast, a deletion that is evenly divisible by three is called an *in-frame* deletion.

Deletions are responsible for an array of genetic disorders, including some cases of male infertility and two thirds of cases of Duchenne muscular dystrophy. Deletion of part of the short arm of chromosome 5 results in Cri du chat syndrome. Deletions in the SMN-encoding gene cause spinal muscular atrophy, the most common genetic cause of infant death.

Recent work suggests that some deletions of highly conserved sequences (CONDELs) may be responsible for the evolutionary differences present among closely related species. Such deletions in humans are referred to as hCONDELs may be responsible for the anatomical and behavioral differences between humans, chimpanzees and other mammals.

Detection

The introduction of molecular techniques in conjunction with classical cytogenetic methods has in recent years greatly improved the diagnostic potential for chromosomal abnormalities. In particular, microarray-comparative genomic hybridization (CGH) based on the use of BAC clones promises a sensitive strategy for the detection of DNA copy-number changes on a genome-wide scale. The resolution of detection could as high as >30,000 "bands" and the size of chromosomal deletion detected could as small as 5–20 kb in length. Other computation methods were selected to discover DNA sequencing deletion errors such as end-sequence profiling.

Mutation

In biology, a mutation is the permanent alteration of the nucleotide sequence of the genome of an organism, virus, or extrachromosomal DNA or other genetic elements. Mutations result from errors during DNA replication or other types of damage to DNA, which then may undergo error-prone repair (especially microhomology-mediated end joining), or cause an error during other forms of repair, or else may cause an error during replication (translesion synthesis). Mutations may also result from insertion or deletion of segments of DNA due to mobile genetic elements. Mutations may or may not produce discernible changes in the observable characteristics (phenotype) of an organism. Mutations play a part in both normal and abnormal biological processes including: evolution, cancer, and the development of the immune system, including junctional diversity.

The genomes of RNA viruses are based on RNA rather than DNA. The RNA viral genome can be double stranded (as in DNA) or single stranded. In some of these viruses (such as the single stranded human immunodeficiency virus) replication occurs quickly and there are no mechanisms to check the genome for accuracy. This error-prone process often results in mutations.

Mutation can result in many different types of change in sequences. Mutations in genes can either have no effect, alter the product of a gene, or prevent the gene from functioning properly or completely. Mutations can also occur in nongenic regions. One study on genetic variations between different species of *Drosophila* suggests that, if a mutation changes a protein produced by a gene, the result is likely to be harmful, with an estimated 70 percent of amino acid polymorphisms that have damaging effects, and the remainder being either neutral or marginally beneficial. Due to the damaging effects that mutations can have on genes, organisms have mechanisms such as DNA repair to prevent or correct mutations by reverting the mutated sequence back to its original state.

Description

Mutations can involve the duplication of large sections of DNA, usually through genetic recombination. These duplications are a major source of raw material for evolving new genes, with tens to hundreds of genes duplicated in animal genomes every million years. Most genes belong to larger gene families of shared ancestry, known as homology. Novel genes are produced by several methods, commonly through the duplication and mutation of an ancestral gene, or by recombining parts of different genes to form new combinations with new functions.

Here, protein domains act as modules, each with a particular and independent function, that can be mixed together to produce genes encoding new proteins with novel properties. For example, the human eye uses four genes to make structures that sense light: three for cone cell or color vision and one for rod cell or night vision; all four arose from a single ancestral gene. Another advantage of duplicating a gene (or even an entire genome) is that this increases engineering redundancy; this allows one gene in the pair to acquire a new function while the other copy performs the original function. Other types of mutation occasionally create new genes from previously noncoding DNA.

Changes in chromosome number may involve even larger mutations, where segments of the DNA within chromosomes break and then rearrange. For example, in the Homininae, two chromosomes fused to produce human chromosome 2; this fusion did not occur in the lineage of the other apes, and they retain these separate chromosomes. In evolution, the most important role of such chromosomal rearrangements may be to accelerate the divergence of a population into new species by making populations less likely to interbreed, thereby preserving genetic differences between these populations.

Sequences of DNA that can move about the genome, such as transposons, make up a major fraction of the genetic material of plants and animals, and may have been important in the evolution of genomes. For example, more than a million copies of the Alu sequence are present in the human genome, and these sequences have now been recruited to perform functions such as regulating gene expression. Another effect of these mobile DNA sequences is that when they move within a genome, they can mutate or delete existing genes and thereby produce genetic diversity.

Nonlethal mutations accumulate within the gene pool and increase the amount of genetic variation. The abundance of some genetic changes within the gene pool can be reduced by natural selection, while other "more favorable" mutations may accumulate and result in adaptive changes.

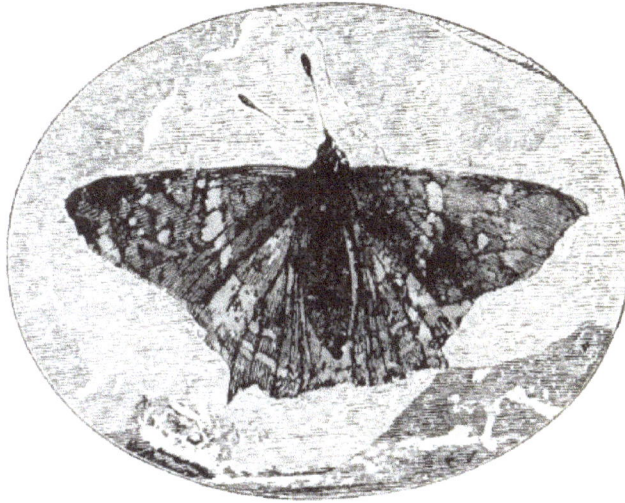

Prodryas persephone, a Late Eocene butterfly

For example, a butterfly may produce offspring with new mutations. The majority of these mutations will have no effect; but one might change the color of one of the butterfly's offspring, making it harder (or easier) for predators to see. If this color change is advantageous, the chance of this butterfly's surviving and producing its own offspring are a little better, and over time the number of butterflies with this mutation may form a larger percentage of the population.

Neutral mutations are defined as mutations whose effects do not influence the fitness of an individual. These can accumulate over time due to genetic drift. It is believed that the overwhelming majority of mutations have no significant effect on an organism's fitness. Also, DNA repair mechanisms are able to mend most changes before they become permanent mutations, and many organisms have mechanisms for eliminating otherwise-permanently mutated somatic cells.

Beneficial mutations can improve reproductive success.

Causes

Four classes of mutations are (1) spontaneous mutations (molecular decay), (2) mutations due to error-prone replication bypass of naturally occurring DNA damage (also called error-prone translesion synthesis), (3) errors introduced during DNA repair, and (4) induced mutations caused by mutagens. Scientists may also deliberately introduce mutant sequences through DNA manipulation for the sake of scientific experimentation.

Spontaneous Mutation

Spontaneous mutations on the molecular level can be caused by:

- Tautomerism — A base is changed by the repositioning of a hydrogen atom, altering the hydrogen bonding pattern of that base, resulting in incorrect base pairing during replication.

- Depurination — Loss of a purine base (A or G) to form an apurinic site (AP site).

- Deamination — Hydrolysis changes a normal base to an atypical base containing a keto group in place of the original amine group. Examples include C → U and A → HX (hypoxanthine), which can be corrected by DNA repair mechanisms; and 5MeC (5-methylcytosine) → T, which is less likely to be detected as a mutation because thymine is a normal DNA base.

- Slipped strand mispairing — Denaturation of the new strand from the template during replication, followed by renaturation in a different spot ("slipping"). This can lead to insertions or deletions.

Error-prone Replication Bypass

There is increasing evidence that the majority of spontaneously arising mutations are due to error-prone replication (translesion synthesis) past DNA damage in the template strand. Naturally occurring oxidative DNA damages arise at least 10,000 times per cell per day in humans and 50,000 times or more per cell per day in rats. In mice, the majority of mutations are caused by translesion synthesis. Likewise, in yeast, Kunz et al. found that more than 60% of the spontaneous single base pair substitutions and deletions were caused by translesion synthesis.

Errors Introduced During DNA Repair

Although naturally occurring double-strand breaks occur at a relatively low frequency in DNA, their repair often causes mutation. Non-homologous end joining (NHEJ) is a major pathway for repairing double-strand breaks. NHEJ involves removal of a few nucleotides to allow somewhat inaccurate alignment of the two ends for rejoining followed by addition of nucleotides to fill in gaps. As a consequence, NHEJ often introduces mutations.

Induced Mutation

Induced mutations on the molecular level can be caused by:-

- Chemicals

    o   Hydroxylamine

    o   Base analogs (e.g., Bromodeoxyuridine (BrdU))

    o   Alkylating agents (e.g., *N*-ethyl-*N*-nitrosourea (ENU)). These agents can mutate both replicating and non-replicating DNA. In contrast, a base analog can mutate the DNA only when the analog is incorporated in replicating the DNA. Each of these classes of chemical mutagens has certain effects that then lead to transitions, transversions, or deletions.

    o   Agents that form DNA adducts (e.g., ochratoxin A)

    o   DNA intercalating agents (e.g., ethidium bromide)

    o   DNA crosslinkers

    o   Oxidative damage

o   Nitrous acid converts amine groups on A and C to diazo groups, altering their hydrogen bonding patterns, which leads to incorrect base pairing during replication.

- Radiation

  o   Ultraviolet light (UV) (non-ionizing radiation). Two nucleotide bases in DNA—cytosine and thymine—are most vulnerable to radiation that can change their properties. UV light can induce adjacent pyrimidine bases in a DNA strand to become covalently joined as a pyrimidine dimer. UV radiation, in particular longer-wave UVA, can also cause oxidative damage to DNA.

Classification of Mutation Types

The sequence of a gene can be altered in a number of ways. Gene mutations have varying effects on health depending on where they occur and whether they alter the function of essential proteins. Mutations in the structure of genes can be classified as:

- Small-scale mutations, such as those affecting a small gene in one or a few nucleotides, including:

  o   Substitution mutations, often caused by chemicals or malfunction of DNA replication, exchange a single nucleotide for another. These changes are classified as transitions or transversions. Most common is the transition that exchanges a purine for a purine (A ↔ G) or a pyrimidine for a pyrimidine, (C ↔ T). A transition can be caused by nitrous acid, base mis-pairing, or mutagenic base analogs such as BrdU. Less common is a transversion, which exchanges a purine for a pyrimidine or a pyrimidine for a purine (C/T ↔ A/G). An example of a transversion is the conversion of adenine (A) into a cytosine (C). A point mutation can be reversed by another point mutation, in which the nucleotide is changed back to its original state (true reversion) or by second-site reversion (a complementary mutation elsewhere that results in regained gene functionality). Point mutations that occur within the protein coding region of a gene may be classified into three kinds, depending upon what the erroneous codon codes for:

    ☐ Silent mutations, which code for the same (or a sufficiently similar) amino acid.

    ☐ Missense mutations, which code for a different amino acid.

    ☐ Nonsense mutations, which code for a stop codon and can truncate the protein.

  o   Insertions add one or more extra nucleotides into the DNA. They are usually caused by transposable elements, or errors during replication of repeating elements. Insertions in the coding region of a gene may alter splicing of the mRNA (splice site mutation), or cause a shift in the reading frame (frameshift), both of which can significantly alter the gene product. Insertions can be reversed by excision of the transposable element.

- o Deletions remove one or more nucleotides from the DNA. Like insertions, these mutations can alter the reading frame of the gene. In general, they are irreversible: Though exactly the same sequence might in theory be restored by an insertion, transposable elements able to revert a very short deletion (say 1–2 bases) in *any* location either are highly unlikely to exist or do not exist at all.

- Large-scale mutations in chromosomal structure, including:

  - o Amplifications (or gene duplications) leading to multiple copies of all chromosomal regions, increasing the dosage of the genes located within them.

  - o Deletions of large chromosomal regions, leading to loss of the genes within those regions.

  - o Mutations whose effect is to juxtapose previously separate pieces of DNA, potentially bringing together separate genes to form functionally distinct fusion genes (e.g., bcr-abl). These include:

    - ☐ Chromosomal translocations: interchange of genetic parts from nonhomologous chromosomes.

    - ☐ Interstitial deletions: an intra-chromosomal deletion that removes a segment of DNA from a single chromosome, thereby apposing previously distant genes. For example, cells isolated from a human astrocytoma, a type of brain tumor, were found to have a chromosomal deletion removing sequences between the Fused in Glioblastoma (FIG) gene and the receptor tyrosine kinase (ROS), producing a fusion protein (FIG-ROS). The abnormal FIG-ROS fusion protein has constitutively active kinase activity that causes oncogenic transformation (a transformation from normal cells to cancer cells).

    - ☐ Chromosomal inversions: reversing the orientation of a chromosomal segment.

  - o Loss of heterozygosity: loss of one allele, either by a deletion or a genetic recombination event, in an organism that previously had two different alleles.

By Effect on Function

- Loss-of-function mutations, also called inactivating mutations, result in the gene product having less or no function (being partially or wholly inactivated). When the allele has a complete loss of function (null allele), it is often called an amorphic mutation in the Muller's morphs schema. Phenotypes associated with such mutations are most often recessive. Exceptions are when the organism is haploid, or when the reduced dosage of a normal gene product is not enough for a normal phenotype (this is called haploinsufficiency).

- Gain-of-function mutations, also called activating mutations, change the gene product such that its effect gets stronger (enhanced activation) or even is superseded by a different and abnormal function. When the new allele is created, a heterozygote containing the newly created allele as well as the original will express the new allele; genetically this defines the mutations as dominant phenotypes. Often called a neomorphic mutation.

- Dominant negative mutations (also called antimorphic mutations) have an altered gene product that acts antagonistically to the wild-type allele. These mutations usually result in an altered molecular function (often inactive) and are characterized by a dominant or semi-dominant phenotype. In humans, dominant negative mutations have been implicated in cancer (e.g., mutations in genes p53, ATM, CEBPA and PPARgamma). Marfan syndrome is caused by mutations in the *FBN1* gene, located on chromosome 15, which encodes fibrillin-1, a glycoprotein component of the extracellular matrix. Marfan syndrome is also an example of dominant negative mutation and haploinsufficiency.

- Lethal mutations are mutations that lead to the death of the organisms that carry the mutations.

- A back mutation or reversion is a point mutation that restores the original sequence and hence the original phenotype.

By Effect on Fitness

In applied genetics, it is usual to speak of mutations as either harmful or beneficial.

- A harmful, or deleterious, mutation decreases the fitness of the organism.

- A beneficial, or advantageous mutation increases the fitness of the organism. Mutations that promotes traits that are desirable, are also called beneficial. In theoretical population genetics, it is more usual to speak of mutations as deleterious or advantageous than harmful or beneficial.

- A neutral mutation has no harmful or beneficial effect on the organism. Such mutations occur at a steady rate, forming the basis for the molecular clock. In the neutral theory of molecular evolution, neutral mutations provide genetic drift as the basis for most variation at the molecular level.

- A nearly neutral mutation is a mutation that may be slightly deleterious or advantageous, although most nearly neutral mutations are slightly deleterious.

Distribution of Fitness Effects

Attempts have been made to infer the distribution of fitness effects (DFE) using mutagenesis experiments and theoretical models applied to molecular sequence data. DFE, as used to determine the relative abundance of different types of mutations (i.e., strongly deleterious, nearly neutral or advantageous), is relevant to many evolutionary questions, such as the maintenance of genetic variation, the rate of genomic decay, the maintenance of outcrossing sexual reproduction as opposed to inbreeding and the evolution of sex and genetic recombination. In summary, the DFE plays an important role in predicting evolutionary dynamics. A variety of approaches have been used to study the DFE, including theoretical, experimental and analytical methods.

- Mutagenesis experiment: The direct method to investigate the DFE is to induce mutations and then measure the mutational fitness effects, which has already been done in viruses, bacteria, yeast, and *Drosophila*. For example, most studies of the DFE in viruses used site-directed mutagenesis to create point mutations and measure relative fitness of each mutant. In *Escherichia coli*, one study used transposon mutagenesis to directly measure the fitness of a random insertion of a derivative of Tn10. In yeast, a combined mutagenesis and deep sequencing approach has been developed to generate high-quality systematic mutant libraries and measure fitness in high throughput. However, given that many mutations have effects too small to be detected and that mutagenesis experiments can detect only mutations of moderately large effect; DNA sequence data analysis can provide valuable information about these mutations.

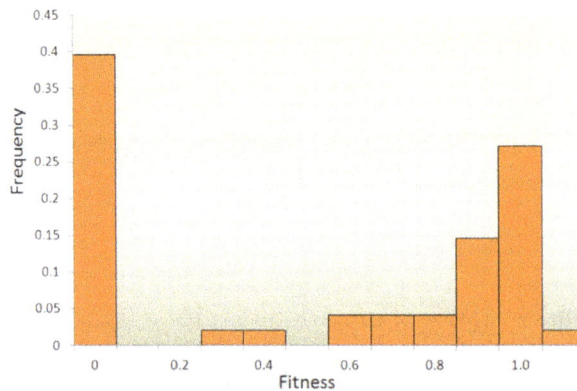

The distribution of fitness effects (DFE) of mutations in vesicular stomatitis virus. In this experiment, random mutations were introduced into the virus by site-directed mutagenesis, and the fitness of each mutant was compared with the ancestral type. A fitness of zero, less than one, one, more than one, respectively, indicates that mutations are lethal, deleterious, neutral, and advantageous.

- Molecular sequence analysis: With rapid development of DNA sequencing technology, an enormous amount of DNA sequence data is available and even more is forthcoming in the future. Various methods have been developed to infer the DFE from DNA sequence data. By examining DNA sequence differences within and between species, we are able to infer various characteristics of the DFE for neutral, deleterious and advantageous mutations. To be specific, the DNA sequence analysis approach allows us to estimate the effects of mutations with very small effects, which are hardly detectable through mutagenesis experiments.

One of the earliest theoretical studies of the distribution of fitness effects was done by Motoo Kimura, an influential theoretical population geneticist. His neutral theory of molecular evolution proposes that most novel mutations will be highly deleterious, with a small fraction being neutral. Hiroshi Akashi more recently proposed a bimodal model for the DFE, with modes centered around highly deleterious and neutral mutations. Both theories agree that the vast majority of novel mutations are neutral or deleterious and that advantageous mutations are rare, which has been supported by experimental results. One example is a study done on the DFE of random mutations in vesicular stomatitis virus. Out of all mutations, 39.6% were lethal, 31.2% were non-lethal deleterious, and 27.1% were neutral. Another example comes from a high throughput mutagenesis experiment with yeast. In this experiment it was shown that the overall DFE is bimodal, with a cluster of neutral mutations, and a broad distribution of deleterious mutations.

Though relatively few mutations are advantageous, those that are play an important role in evolutionary changes. Like neutral mutations, weakly selected advantageous mutations can be lost due to random genetic drift, but strongly selected advantageous mutations are more likely to be fixed. Knowing the DFE of advantageous mutations may lead to increased ability to predict the evolutionary dynamics. Theoretical work on the DFE for advantageous mutations has been done by John H. Gillespie and H. Allen Orr. They proposed that the distribution for advantageous mutations should be exponential under a wide range of conditions, which, in general, has been supported by experimental studies, at least for strongly selected advantageous mutations.

In general, it is accepted that the majority of mutations are neutral or deleterious, with rare mutations being advantageous; however, the proportion of types of mutations varies between species. This indicates two important points: first, the proportion of effectively neutral mutations is likely to vary between species, resulting from dependence on effective population size; second, the average effect of deleterious mutations varies dramatically between species. In addition, the DFE also differs between coding regions and noncoding regions, with the DFE of noncoding DNA containing more weakly selected mutations.

By Impact on Protein Sequence

- A frameshift mutation is a mutation caused by insertion or deletion of a number of nucleotides that is not evenly divisible by three from a DNA sequence. Due to the triplet nature of gene expression by codons, the insertion or deletion can disrupt the reading frame, or the grouping of the codons, resulting in a completely different translation from the original. The earlier in the sequence the deletion or insertion occurs, the more altered the protein produced is. In contrast, any insertion or deletion that is evenly divisible by three is termed an *in-frame mutation*

- A nonsense mutation is a point mutation in a sequence of DNA that results in a premature stop codon, or a *nonsense codon* in the transcribed mRNA, and possibly a truncated, and often nonfunctional protein product.

- Missense mutations or *nonsynonymous mutations* are types of point mutations where a single nucleotide is changed to cause substitution of a different amino acid. This in turn can render the resulting protein nonfunctional. Such mutations are responsible for diseases such as Epidermolysis bullosa, sickle-cell disease, and SOD1-mediated ALS.

- A neutral mutation is a mutation that occurs in an amino acid codon that results in the use of a different, but chemically similar, amino acid. The similarity between the two is enough that little or no change is often rendered in the protein. For example, a change from AAA to AGA will encode arginine, a chemically similar molecule to the intended lysine.

- Silent mutations are mutations that do not result in a change to the amino acid sequence of a protein, unless the changed amino acid is sufficiently similar to the original. They may occur in a region that does not code for a protein, or they may occur within a codon in a manner that does not alter the final amino acid sequence. The phrase *silent mutation* is often used interchangeably with the phrase *synonymous mutation*; however, synonymous mutations are a subcategory of the former, occurring only within exons (and necessarily exactly preserving the amino acid sequence of the protein). Synonymous mutations occur due to the degenerate nature of the genetic code.

By Inheritance

A mutation has caused this garden moss rose to produce flowers of different colors. This is a somatic mutation that may also be passed on in the germline.

In multicellular organisms with dedicated reproductive cells, mutations can be subdivided into germline mutations, which can be passed on to descendants through their reproductive cells, and somatic mutations (also called acquired mutations), which involve cells outside the dedicated reproductive group and which are not usually transmitted to descendants.

A germline mutation gives rise to a *constitutional mutation* in the offspring, that is, a mutation that is present in every cell. A constitutional mutation can also occur very soon after fertilisation, or continue from a previous constitutional mutation in a parent.

The distinction between germline and somatic mutations is important in animals that have a dedicated germline to produce reproductive cells. However, it is of little value in understanding the effects of mutations in plants, which lack dedicated germline. The distinction is also blurred in those animals that reproduce asexually through mechanisms such as budding, because the cells that give rise to the daughter organisms also give rise to that organism's germline. A new germline mutation that was not inherited from either parent is called a *de novo* mutation.

Diploid organisms (e.g., humans) contain two copies of each gene—a paternal and a maternal allele. Based on the occurrence of mutation on each chromosome, we may classify mutations into three types.

- A heterozygous mutation is a mutation of only one allele.

- A homozygous mutation is an identical mutation of both the paternal and maternal alleles.

- Compound heterozygous mutations or a genetic compound comprises two different mutations in the paternal and maternal alleles.

A wild type or homozygous non-mutated organism is one in which neither allele is mutated.

Special Classes

- Conditional mutation is a mutation that has wild-type (or less severe) phenotype under certain "permissive" environmental conditions and a mutant phenotype under certain "restrictive" conditions. For example, a temperature-sensitive mutation can cause cell death at high temperature (restrictive condition), but might have no deleterious consequences at a lower temperature (permissive condition).

- Replication timing quantitative trait loci affects DNA replication.

Nomenclature

In order to categorize a mutation as such, the "normal" sequence must be obtained from the DNA of a "normal" or "healthy" organism (as opposed to a "mutant" or "sick" one), it should be identified and reported; ideally, it should be made publicly available for a straightforward nucleotide-by-nucleotide comparison, and agreed upon by the scientific community or by a group of expert geneticists and biologists, who have the responsibility of establishing the *standard* or so-called "consen-sus" sequence. This step requires a tremendous scientific effort. Once the consensus sequence is known, the mutations in a genome can be pinpointed, described, and classified. The committee of the Human Genome Variation Society (HGVS) has developed the standard human sequence variant nomenclature, which should be used by researchers and DNA diagnostic centers to generate unambiguous mutation descriptions. In principle, this nomenclature can also be used to describe mutations in other organisms. The nomenclature specifies the type of mutation and base or amino acid changes.

- Nucleotide substitution (e.g., 76A>T) — The number is the position of the nucleotide from the 5' end; the first letter represents the wild-type nucleotide, and the second letter represents the nucleotide that replaced the wild type. In the given example, the adenine at the 76th position was replaced by a thymine.

    o   If it becomes necessary to differentiate between mutations in genomic DNA, mitochondrial DNA, and RNA, a simple convention is used. For example, if the 100th base of a nucleotide sequence mutated from G to C, then it would be written as g.100G>C if the mutation occurred in genomic DNA, m.100G>C if the mutation occurred in mitochondrial DNA, or r.100g>c if the mutation occurred in RNA. Note that, for mutations in RNA, the nucleotide code is written in lower case.

- Amino acid substitution (e.g., D111E) — The first letter is the one letter code of the wild-type amino acid, the number is the position of the amino acid from the N-terminus, and the second letter is the one letter code of the amino acid present in the mutation. Nonsense mutations are represented with an X for the second amino acid (e.g. D111X).

- Amino acid deletion (e.g., ΔF508) — The Greek letter Δ (delta) indicates a deletion. The letter refers to the amino acid present in the wild type and the number is the position from the N terminus of the amino acid were it to be present as in the wild type.

Mutation Rates

Mutation rates vary substantially across species, and the evolutionary forces that generally determine mutation are the subject of ongoing investigation.

Harmful Mutations

Changes in DNA caused by mutation can cause errors in protein sequence, creating partially or completely non-functional proteins. Each cell, in order to function correctly, depends on thousands of proteins to function in the right places at the right times. When a mutation alters a protein that plays a critical role in the body, a medical condition can result. A condition caused by mutations in one or more genes is called a genetic disorder. Some mutations alter a gene's DNA base sequence but do not change the function of the protein made by the gene. One study on the comparison of genes between different species of *Drosophila* suggests that if a mutation does change a protein, this will probably be harmful, with an estimated 70 percent of amino acid polymorphisms having damaging effects, and the remainder being either neutral or weakly beneficial. Studies have shown that only 7% of point mutations in noncoding DNA of yeast are deleterious and 12% in coding DNA are deleterious. The rest of the mutations are either neutral or slightly beneficial.

If a mutation is present in a germ cell, it can give rise to offspring that carries the mutation in all of its cells. This is the case in hereditary diseases. In particular, if there is a mutation in a DNA repair gene within a germ cell, humans carrying such germline mutations may have an increased risk of cancer. A list of 34 such germline mutations is given in the article DNA repair-deficiency disorder. An example of one is albinism, a mutation that occurs in the OCA1 or OCA2 gene. Individuals with this disorder are more prone to many types of cancers, other disorders and have impaired vision. On the other hand, a mutation may occur in a somatic cell of an organism. Such mutations will be present in all descendants of this cell within the same organism, and certain mutations can cause the cell to become malignant, and, thus, cause cancer.

A DNA damage can cause an error when the DNA is replicated, and this error of replication can cause a gene mutation that, in turn, could cause a genetic disorder. DNA damages are repaired by the DNA repair system of the cell. Each cell has a number of pathways through which enzymes recognize and repair damages in DNA. Because DNA can be damaged in many ways, the process of DNA repair is an important way in which the body protects itself from disease. Once DNA damage has given rise to a mutation, the mutation cannot be repaired. DNA repair pathways can only recognize and act on "abnormal" structures in the DNA. Once a mutation occurs in a gene sequence it then has normal DNA structure and cannot be repaired.

Beneficial Mutations

Although mutations that cause changes in protein sequences can be harmful to an organism, on occasions the effect may be positive in a given environment. In this case, the mutation may enable the mutant organism to withstand particular environmental stresses better than wild-type organisms, or reproduce more quickly. In these cases a mutation will tend to become more common in a population through natural selection.

For example, a specific 32 base pair deletion in human CCR5 (CCR5-Δ32) confers HIV resistance to homozygotes and delays AIDS onset in heterozygotes. One possible explanation of the etiology of the relatively high frequency of CCR5-Δ32 in the European population is that it conferred resistance to the bubonic plague in mid-14th century Europe. People with this mutation were more likely to survive infection; thus its frequency in the population increased. This theory could explain why this mutation is not found in Southern Africa, which remained untouched by bubonic plague. A newer theory suggests that the selective pressure on the CCR5 Delta 32 mutation was caused by smallpox instead of the bubonic plague.

Another example is sickle-cell disease, a blood disorder in which the body produces an abnormal type of the oxygen-carrying substance hemoglobin in the red blood cells. One-third of all indigenous inhabitants of Sub-Saharan Africa carry the gene, because, in areas where malaria is common, there is a survival value in carrying only a single sickle-cell gene (sickle cell trait). Those with only one of the two alleles of the sickle-cell disease are more resistant to malaria, since the infestation of the malaria *Plasmodium* is halted by the sickling of the cells that it infests.

Prion Mutations

Prions are proteins and do not contain genetic material. However, prion replication has been shown to be subject to mutation and natural selection just like other forms of replication. The human gene PRNP codes for the major prion protein, PrP, and is subject to mutations that can give rise to disease-causing prions.

Somatic Mutations

A change in the genetic structure that is not inherited from a parent, and also not passed to offspring, is called a *somatic cell genetic mutation* or *acquired mutation*.

Cells with heterozygous mutations (one good copy of gene and one mutated copy) may function normally with the unmutated copy until the good copy has been spontaneously somatically mutated. This kind of mutation happens all the time in living organisms, but it is difficult to measure the rate. Measuring this rate is important in predicting the rate at which people may develop cancer.

Point mutations may arise from spontaneous mutations that occur during DNA replication. The rate of mutation may be increased by mutagens. Mutagens can be physical, such as radiation from UV rays, X-rays or extreme heat, or chemical (molecules that misplace base pairs or disrupt the helical shape of DNA). Mutagens associated with cancers are often studied to learn about cancer and its prevention.

Natural Selection

Natural selection is the differential survival and reproduction of individuals due to differences in phenotype. It is a key mechanism of evolution, the change in heritable traits of a population over time. Charles Darwin popularised the term "natural selection"; he compared it with artificial selection (selective breeding).

Variation exists within all populations of organisms. This occurs partly because random mutations arise in the genome of an individual organism, and offspring can inherit such mutations. Throughout the lives of the individuals, their genomes interact with their environments to cause variations in traits. (The environment of a genome includes the molecular biology in the cell, other cells, other individuals, populations, species, as well as the abiotic environment.) Individuals with certain variants of the trait may survive and reproduce more than individuals with other, less successful, variants. Therefore, the population evolves. Factors that affect reproductive success are also important, an issue that Darwin developed in his ideas on sexual selection (now often included in natural selection) and on fecundity selection, for example.

Natural selection acts on the phenotype, or the observable characteristics of an organism, but the genetic (heritable) basis of any phenotype that gives a reproductive advantage may become more common in a population. Over time, this process can result in populations that specialise for particular ecological niches (microevolution) and may eventually result in the emergence of new species (macroevolution). In other words, natural selection is an important process (though not the only process) by which evolution takes place within a population of organisms. Natural selection can be contrasted with artificial selection, in which humans intentionally choose specific traits (although they may not always get what they want). In natural selection there is no intentional choice. In other words, artificial selection is teleological and natural selection is not teleological, though biologists often use teleological language to describe it.

Natural selection is one of the cornerstones of modern biology. The concept, published by Darwin and Alfred Russel Wallace in a joint presentation of papers in 1858, was elaborated in Darwin's influential 1859 book *On the Origin of Species*, which described natural selection as analogous to artificial selection, a process by which animals and plants with traits considered desirable by human breeders are systematically favoured for reproduction. The concept of natural selection originally developed in the absence of a valid theory of heredity; at the time of Darwin's writing, science had yet to develop modern theories of genetics. The union of traditional Darwinian evolution with subsequent discoveries in classical and molecular genetics is termed the *modern evolutionary synthesis*. Natural selection remains the primary explanation for adaptive evolution.

Natural variation occurs among the individuals of any population of organisms. Many of these differences do not affect survival or reproduction, but some differences may improve the chances of survival and reproduction of a particular individual. A rabbit that runs faster than others may be more likely to escape from predators, and algae that are more efficient at extracting energy from sunlight will grow faster. Something that increases an organism's chances of survival will often also include its reproductive rate; however, sometimes there is a trade-off between survival and current reproduction. Ultimately, what matters is total lifetime reproduction of the organism.

The peppered moth exists in both light and dark colours in the United Kingdom, but during the industrial revolution, many of the trees on which the moths rested became blackened by soot, giving the dark-coloured moths an advantage in hiding from predators. This gave dark-coloured moths a better chance of surviving to produce dark-coloured offspring, and in just fifty years from the first dark moth being caught, nearly all of the moths in industrial Manchester were dark. The balance was reversed by the effect of the Clean Air Act 1956, and the dark moths became rare again, demonstrating the influence of natural selection on peppered moth evolution.

If the traits that give these individuals a reproductive advantage are also heritable, that is, passed from parent to offspring, then there will be a slightly higher proportion of fast rabbits or efficient algae in the next generation. This is known as *differential reproduction*. Even if the reproductive advantage is very slight, over many generations any heritable advantage will become dominant in the population. In this way the natural environment of an organism "selects" for traits that confer a reproductive advantage, causing gradual changes or evolution of life. This effect was first described and named by Charles Darwin.

The concept of natural selection predates the understanding of genetics, the mechanism of heredity for all known life forms. In modern terms, selection acts on an organism's phenotype, or observable characteristics, but it is the organism's genetic make-up or genotype that is inherited. The phenotype is the result of the genotype and the environment in which the organism lives.

This is the link between natural selection and genetics, as described in the modern evolutionary synthesis. Although a complete theory of evolution also requires an account of how genetic variation arises in the first place (such as by mutation and sexual reproduction) and includes other evolutionary mechanisms (such as genetic drift and gene flow), natural selection appears to be the most important mechanism for creating complex adaptations in nature.

Nomenclature and Usage

The term *natural selection* has slightly different definitions in different contexts. It is most often defined to operate on heritable traits, because these are the traits that directly participate in evolution. However, natural selection is "blind" in the sense that changes in phenotype (physical and behavioural characteristics) can give a reproductive advantage regardless of whether or not the trait is heritable (non heritable traits can be the result of environmental factors or the life experience of the organism).

Following Darwin's primary usage the term is often used to refer to both the evolutionary consequence of blind selection and to its mechanisms. It is sometimes helpful to explicitly distinguish between selection's mechanisms and its effects; when this distinction is important, scientists define "natural selection" specifically as "those mechanisms that contribute to the selection of individuals that reproduce," without regard to whether the basis of the selection is heritable. This is sometimes referred to as "phenotypic natural selection."

Traits that cause greater reproductive success of an organism are said to be selected for, whereas those that reduce success are selected against. Selection for a trait may also result in the selection of other correlated traits that do not themselves directly influence reproductive advantage. This may occur as a result of pleiotropy or gene linkage.

Fitness

The concept of fitness is central to natural selection. In broad terms, individuals that are more "fit" have better potential for survival, as in the well-known phrase "survival of the fittest." However, as with natural selection above, the precise meaning of the term is much more subtle. Modern evolutionary theory defines fitness not by how long an organism lives, but by how successful it is at reproducing. If an organism lives half as long as others of its species, but has twice as many offspring surviving to adulthood, its genes will become more common in the adult population of the next generation.

1. Geospiza magnirostris
3. Geospiza parvula
2. Geospiza fortis
4. Certhidea olivacea

Finches from Galapagos Archipelago

Charles Darwin's illustrations of beak variation in the finches of the Galápagos Islands, which hold 13 closely related species that differ most markedly in the shape of their beaks. The beak of each species is suited to its preferred food, suggesting that beak shapes evolved by natural selection.

Though natural selection acts on individuals, the effects of chance mean that fitness can only really be defined "on average" for the individuals within a population. The fitness of a particular genotype corresponds to the average effect on all individuals with that genotype. Very low-fitness genotypes cause their bearers to have few or no offspring on average; examples include many human genetic disorders like cystic fibrosis.

Since fitness is an averaged quantity, it is also possible that a favourable mutation arises in an individual that does not survive to adulthood for unrelated reasons. Fitness also depends crucially upon the environment. Conditions like sickle-cell anaemia may have low fitness in the general human population, but because the sickle cell trait confers immunity from malaria, it has high fitness value in populations that have high malaria infection rates.

Types of Selection

Natural selection can act on any heritable phenotypic trait, and selective pressure can be produced by any aspect of the environment, including sexual selection and competition with members of the same or other species. However, this does not imply that natural selection is always directional and results in adaptive evolution; natural selection often results in the maintenance of the status quo by eliminating less fit variants.

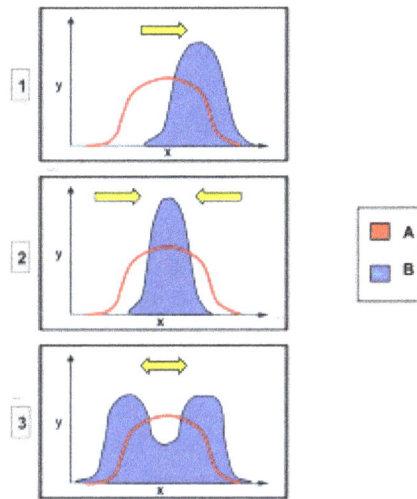

These charts depict the different types of genetic selection. On each graph, the x-axis variable is the type of pheno-typic trait and the y-axis variable is the amount of organisms. Group A is the original population and Group B is the population after selection. Graph 1 shows directional selection, in which a single extreme phenotype is favored. Graph 2 depicts stabilizing selection, where the intermediate phenotype is favored over the extreme traits. Graph 3 shows disruptive selection, in which the extreme phenotypes are favored over the intermediate.

Selection can be classified according to its effect on a trait. Stabilizing selection acts to hold a trait at a stable optimum, and in the simplest case all deviations from this optimum are selectively dis-advantageous. Directional selection acts during transition periods, when the current mode of the trait is sub-optimal, and alters the trait towards a single new optimum. Disruptive selection also acts during transition periods when the current mode is sub-optimal, but alters the trait in more than one direction. In particular, if the trait is quantitative and univariate then both higher and lower trait levels are favoured. Disruptive selection can be a precursor to speciation.

Selection can also be classified according to its effect on allele frequency. Positive selection acts to increase the frequency of an allele. Negative selection acts to decrease the frequency of an allele. Note that for a diallelic locus, positive selection on one allele perforce implies negative selection on the other allele.

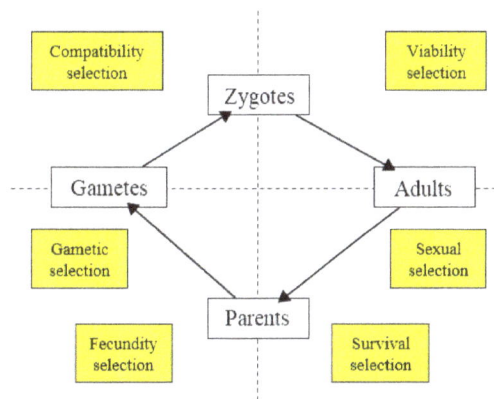

The life cycle of a sexually reproducing organism. Various components of natural selection are indicated for each life stage.

Selection can also be classified according to its effect on genetic diversity. Purifying selection acts to remove genetic variation from the population (and is opposed by *de novo* mutation, which introduces new variation). Balancing selection acts to maintain genetic variation in a population (even in the absence of *de novo* mutation). Mechanisms include negative frequency-dependent selection (of which heterozygote advantage is a special case), and spatial and/or temporal fluctuations in the strength and direction of selection.

Selection can also be classified according to the stage of an organism's life cycle at which it acts. The use of terminology differs here. Some recognise just two types of selection: viability selection (or survival selection) which acts to improve the probability of survival of the organism, and fecundity selection (or fertility selection, or reproductive selection) which acts to improve the rate of reproduction, given successful survival. Others split the life cycle into further components of selection. Thus viability and survival selection may be defined separately and respectively as acting to improve the probability of survival before and after reproductive age is reached, while fecundity selection may be split into additional sub-components including sexual selection, *gametic selection* (acting on gamete survival) and *compatibility selection* (acting on zygote formation).

Selection can also be classified according to the level or unit of selection. *Individual selection* acts at the level of the individual, in the sense that adaptions are 'for' the benefit of the individual, and result from selection among individuals. Gene selection acts directly at the level of the gene. In many situations, this is simply a different way of describing individual selection. However, in some cases (e.g., kin selection and intragenomic conflict), gene-level selection provides a more apt explanation of the underlying process. Group selection acts at the level of groups of organisms. The mechanism assumes that groups replicate and mutate in an analogous way to genes and individuals. There is an ongoing debate over the degree to which group selection occurs in nature.

Finally, selection can be classified according to the resource being competed for. Sexual selection results from competition for mates. Sexual selection can be *intrasexual*, as in cases of competition among individuals of the same sex in a population, or *intersexual*, as in cases where one sex controls reproductive access by choosing among a population of available mates. Typically, sexual selection proceeds via fecundity selection, sometimes at the expense of viability. Ecological selection is natural selection via any other means than sexual selection. Alternatively, natural selection is sometimes defined as synonymous with ecological selection, and sexual selection is then classified as a separate mechanism to natural selection. This accords with Darwin's usage of these terms, but ignores the fact that mate competition and mate choice are natural processes.

Note that types of selection often act in concert. Thus Stabilizing selection typically proceeds via negative selection on rare alleles, leading to purifying selection, while directional selection typically proceeds via positive selection on an initially rare favoured allele.

Ecological selection covers any mechanism of selection as a result of the environment, including relatives (e.g., kin selection, competition, and infanticide).

Sexual Selection

Sexual selection refers specifically to competition for mates, which can be *intrasexual*, between individuals of the same sex, that is male–male competition, or *intersexual*, where one gender choose mates. However, some species exhibit sex-role reversed behaviour in which it is males that are most selective in mate choice; such as in some fishes of the family Syngnathidae, though likely examples have also been found in sexual selection in amphibians, sexual selection in birds, sexual selection in mammals (including sexual selection in humans) and sexual selection in scaled reptiles.

The peacock tail in flight, a classic example of sexual selection.

Alligator Pipefish, a syngnathid

Phenotypic traits can be displayed in one sex and desired in the other sex, causing a positive feedback loop called a Fisherian runaway, for example, the extravagant plumage of some male birds. An alternate theory proposed by the same Ronald Fisher in 1930 is the sexy son hypothesis, that mothers will want promiscuous sons to give them large numbers of grandchildren and so will choose promiscuous fathers for their children. Aggression between members of the same sex is sometimes associated with very distinctive features, such as the antlers of stags, which are used in combat with other stags. More generally, intrasexual selection is often associated with sexual dimorphism, including differences in body size between males and females of a species.

Examples of Natural Selection

A well-known example of natural selection in action is the development of antibiotic resistance in microorganisms. Since the discovery of penicillin in 1928, antibiotics have been used to fight bacterial diseases. Natural populations of bacteria contain, among their vast numbers of individual members, considerable variation in their genetic material, primarily as the result of mutations. When exposed to antibiotics, most bacteria die quickly, but some may have mutations that make them slightly less susceptible. If the exposure to antibiotics is short, these individuals will survive the treatment. This selective elimination of maladapted individuals from a population is natural selection.

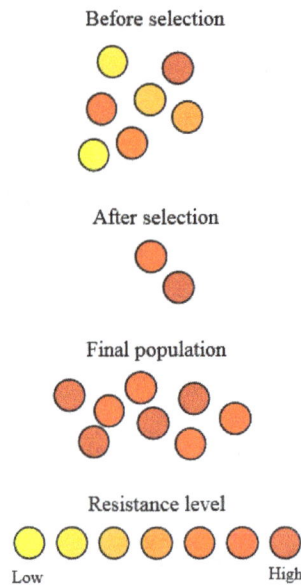

Resistance to antibiotics is increased though the survival of individuals that are immune to the effects of the antibiotic, whose offspring then inherit the resistance, creating a new population of resistant bacteria.

These surviving bacteria will then reproduce again, producing the next generation. Due to the elimination of the maladapted individuals in the past generation, this population contains more bacteria that have some resistance against the antibiotic. At the same time, new mutations occur, contributing new genetic variation to the existing genetic variation. Spontaneous mutations are very rare, and advantageous mutations are even rarer. However, populations of bacteria are large enough that a few individuals will have beneficial mutations. If a new mutation reduces their susceptibility to an antibiotic, these individuals are more likely to survive when next confronted with that antibiotic.

Given enough time and repeated exposure to the antibiotic, a population of antibiotic-resistant bacteria will emerge. This new changed population of antibiotic-resistant bacteria is optimally adapted to the context it evolved in. At the same time, it is not necessarily optimally adapted any more to the old antibiotic free environment. The end result of natural selection is two populations that are both optimally adapted to their specific environment, while both perform substandard in the other environment.

The widespread use and misuse of antibiotics has resulted in increased microbial resistance to antibiotics in clinical use, to the point that the methicillin-resistant *Staphylococcus aureus* (MRSA) has been described as a "superbug" because of the threat it poses to health and its relative invulnerability to existing drugs. Response strategies typically include the use of different, stronger antibiotics; however, new strains of MRSA have recently emerged that are resistant even to these drugs.

This is an example of what is known as an evolutionary arms race, in which bacteria continue to develop strains that are less susceptible to antibiotics, while medical researchers continue to develop new antibiotics that can kill them. A similar situation occurs with pesticide resistance in plants and insects. Arms races are not necessarily induced by man; a well-documented example involves the spread of a gene in the butterfly *Hypolimnas bolina* suppressing male-killing activity by *Wolbachia* bacteria parasites on the island of Samoa, where the spread of the gene is known to have occurred over a period of just five years

Evolution by Means of Natural Selection

A prerequisite for natural selection to result in adaptive evolution, novel traits and speciation, is the presence of heritable genetic variation that results in fitness differences. Genetic variation is the result of mutations, genetic recombinations and alterations in the karyotype (the number, shape, size and internal arrangement of the chromosomes). Any of these changes might have an effect that is highly advantageous or highly disadvantageous, but large effects are very rare. In the past, most changes in the genetic material were considered neutral or close to neutral because they occurred in noncoding DNA or resulted in a synonymous substitution. However, recent research suggests that many mutations in non-coding DNA do have slight deleterious effects. Although both mutation rates and average fitness effects of mutations are dependent on the organism, estimates from data in humans have found that a majority of mutations are slightly deleterious.

The exuberant tail of the peacock is thought to be the result of sexual selection by females. This peacock is leucistic; selection against leucism and albinism in nature is intense because they are easily spotted by predators or are unsuccessful in competition for mates.

By the definition of fitness, individuals with greater fitness are more likely to contribute offspring to the next generation, while individuals with lesser fitness are more likely to die early or fail to reproduce. As a result, alleles that on average result in greater fitness become more abundant in the next generation, while alleles that in general reduce fitness become rarer. If the selection forces remain the same for many generations, beneficial alleles become more and more abundant, until they dominate the population, while alleles with a lesser fitness disappear. In every generation, new mutations and re-combinations arise spontaneously, producing a new spectrum of phenotypes. Therefore, each new generation will be enriched by the increasing abundance of alleles that contribute to those traits that were favoured by selection, enhancing these traits over successive generations.

Some mutations occur in so-called regulatory genes. Changes in these can have large effects on the phenotype of the individual because they regulate the function of many other genes. Most, but not all, mutations in regulatory genes result in non-viable zygotes. Examples of nonlethal regulatory mutations occur in HOX genes in humans, which can result in a cervical rib or polydactyly, an increase in the number of fingers or toes. When such mutations result in a higher fitness, natural selection will favour these phenotypes and the novel trait will spread in the population.

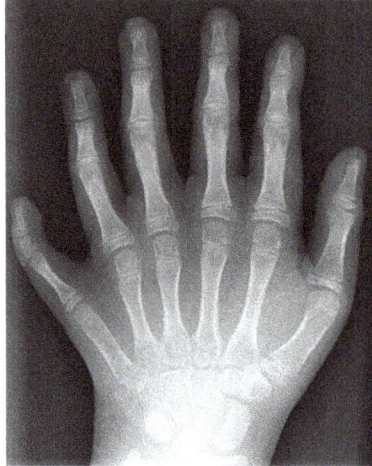

X-ray of the left hand of a ten-year-old boy with polydactyly.

Established traits are not immutable; traits that have high fitness in one environmental context may be much less fit if environmental conditions change. In the absence of natural selection to preserve such a trait, it will become more variable and deteriorate over time, possibly resulting in a vestigial manifestation of the trait, also called evolutionary baggage. In many circumstances, the apparently vestigial structure may retain a limited functionality, or may be co-opted for other advantageous traits in a phenomenon known as preadaptation. A famous example of a vestigial structure, the eye of the blind mole-rat, is believed to retain function in photoperiod perception.

Speciation

Speciation requires cessation of gene flow, which results from reproductive isolation. Over time, isolated subgroups might diverge radically to become different species, either because of differences in selection pressures on the different subgroups, or because different mutations arise spontaneously in the different populations, or because of genetic drift which is responsible for phenomena such as bottleneck effect and founder effect. A lesser-known mechanism of speciation occurs via hybridisation, well-documented in plants and occasionally observed in species-rich groups of animals such as cichlid fishes. Such mechanisms of rapid speciation can reflect a mechanism of evolutionary change known as punctuated equilibrium, which suggests that evolutionary change and in particular speciation typically happens quickly after interrupting long periods of stasis.

Genetic changes within groups result in increasing incompatibility between the genomes of the two subgroups, thus reducing gene flow between the groups. Gene flow will effectively cease when the distinctive mutations characterising each subgroup become fixed. As few as two mutations can result in speciation: if each mutation has a neutral or positive effect on fitness when they occur separately, but a negative effect when they occur together, then fixation of these genes in the respective subgroups will lead to two reproductively isolated populations. According to the biological species concept, these will be two different species.

Historical Development

The modern theory of natural selection derives from the work of Charles Darwin in the nineteenth century.

Pre-darwinian Theories

Several ancient philosophers expressed the idea that nature produces a huge variety of creatures, randomly, and that only those creatures that manage to provide for themselves and reproduce successfully survive; well-known examples include Empedocles and his intellectual successor, the Roman poet Lucretius. Empedocles' idea that organisms arose entirely by the incidental workings of causes such as heat and cold was criticised by Aristotle in Book II of *Physics*. He posited natural teleology in its place. He believed that form was achieved for a purpose, citing the regularity of heredity in species as proof. Nevertheless, he acceded that new types of animals, monstrosities, can occur in very rare instances (*Generation of Animals*, Book IV). As quoted in Darwin's *The Origin of Species* (1872), Aristotle considered whether different forms (e.g., of teeth) might have appeared accidentally, but only the useful forms survived:

So what hinders the different parts [of the body] from having this merely accidental relation in nature? as the teeth, for example, grow by necessity, the front ones sharp, adapted for dividing, and the grinders flat, and serviceable for masticating the food; since they were not made for the sake of this, but it was the result of accident. And in like manner as to the other parts in which there appears to exist an adaptation to an end. Wheresoever, therefore, all things together (that is all the parts of one whole) happened like as if they were made for the sake of something, these were preserved, having been appropriately constituted by an internal spontaneity, and whatsoever things were not thus constituted, perished, and still perish.

—*Aristotle, Physics, Book II, Chapter 8*

But he rejected this possibility in the next paragraph:

...Yet it is impossible that this should be the true view. For teeth and all other natural things either invariably or normally come about in a given way; but of not one of the results of chance or spontaneity is this true. We do not ascribe to chance or mere coincidence the frequency of rain in winter, but frequent rain in summer we do; nor heat in the dog-days, but only if we have it in winter. If then, it is agreed that things are either the result of coincidence or for an end, and these cannot be the result of coincidence or spontaneity, it follows that they must be for an end; and that such things are all due to nature even the champions of the theory which is before us would agree. Therefore action for an end is present in things which come to be and are by nature.

—Aristotle, Physics, Book II, Chapter 8

The struggle for existence was later described by Islamic writer Al-Jahiz in the 9th century.

The classical arguments were reintroduced in the 18th century by Pierre Louis Maupertuis and others, including Charles Darwin's grandfather Erasmus Darwin. While these forerunners had an influence on Darwinism, they later had little influence on the trajectory of evolutionary thought after Charles Darwin.

Until the early 19th century, the prevailing view in Western societies was that differences between individuals of a species were uninteresting departures from their Platonic idealism (or typus) of created kinds. However, the theory of uniformitarianism in geology promoted the idea that simple, weak forces could act continuously over long periods of time to produce radical changes in the Earth's landscape. The success of this theory raised awareness of the vast scale of geological time and made plausible the idea that tiny, virtually imperceptible changes in successive generations could produce consequences on the scale of differences between species.

Early 19th-century evolutionists such as Jean-Baptiste Lamarck suggested the inheritance of acquired characteristics as a mechanism for evolutionary change; adaptive traits acquired by an organism during its lifetime could be inherited by that organism's progeny, eventually causing transmutation of species. This theory has come to be known as Lamarckism and was an influence on the anti-genetic ideas of the Stalinist Soviet biologist Trofim Lysenko.

Between 1835 and 1837, zoologist Edward Blyth also contributed specifically to the area of variation, artificial selection, and how a similar process occurs in nature (Edward Blyth#On natural selection). In fact, Charles Darwin showed his high regards for Blyth's ideas in the first chapter on variation of *On the Origin of Species* that he wrote, "Mr. Blyth, whose opinion, from his large and varied stores of knowledge, I should value more than that of almost any one, ..."

Darwin's Theory

In 1859, Charles Darwin set out his theory of evolution by natural selection as an explanation for adaptation and speciation. He defined natural selection as the "principle by which each slight variation [of a trait], if useful, is preserved." The concept was simple but powerful: individuals best adapted to their environments are more likely to survive and reproduce. As long as there is some variation between them and that variation is heritable, there will be an inevitable selection of individuals with the most advantageous variations. If the variations are inherited, then differential reproductive success will lead to a progressive evolution of particular populations of a species, and populations that evolve to be sufficiently different eventually become different species.

Darwin's ideas were inspired by the observations that he had made on the second voyage of HMS *Beagle* (1831–1836), and by the work of a political economist, the Reverend Thomas Robert Malthus, who in *An Essay on the Principle of Population* (1798), noted that population (if unchecked) increases exponentially, whereas the food supply grows only arithmetically; thus, inevitable limitations of resources would have demographic implications, leading to a "struggle for existence." When Darwin read Malthus in 1838 he was already primed by his work as a naturalist to appreciate the "struggle for existence" in nature and it struck him that as population outgrew resources, "favourable variations would tend to be preserved, and unfavourable ones to be destroyed. The result of this would be the formation of new species."

Here is Darwin's own summary of the idea, which can be found in the fourth chapter of *On the Origin of Species*:

> If during the long course of ages and under varying conditions of life, organic beings vary at all in the several parts of their organisation, and I think this cannot be disputed; if there be, owing to the high geometrical powers of increase of each species, at some age, season, or year, a severe struggle for life, and this certainly cannot be disputed; then, considering the infinite complexity of the relations of all organic beings to each other and to their conditions of existence, causing an infinite diversity in structure, constitution, and habits, to be advantageous to them, I think it would be a most extraordinary fact if no variation ever had occurred useful to each being's own welfare, in the same way as so many variations have occurred useful to man. But if variations useful to any organic being do occur, assuredly individuals thus characterised will have the best chance of being preserved in the struggle for life; and from the strong principle of inheritance they will tend to produce offspring similarly characterised. This principle of preservation, I have called, for the sake of brevity, Natural Selection.

Once he had his theory "by which to work," Darwin was meticulous about gathering and refining evidence as his "prime hobby" before making his idea public. He was in the process of writing his "big book" to present his researches when the naturalist Alfred Russel Wallace independently conceived of the principle and described it in an essay he sent to Darwin to forward to Charles Lyell. Lyell and Joseph Dalton Hooker decided (without Wallace's knowledge) to present his essay together with unpublished writings that Darwin had sent to fellow naturalists, and *On the Tendency of Species to form Varieties; and on the Perpetuation of Varieties and Species by Natural Means of Selection* was read to the Linnean Society of London announcing co-discovery of the principle in July 1858. Darwin published a detailed account of his evidence and conclusions in *On the Origin of Species* in 1859. In the 3rd edition of 1861 Darwin acknowledged that others—a notable one being William Charles Wells in 1813, and Patrick Matthew in 1831—had proposed similar ideas, but had neither developed them nor presented them in notable scientific publications.

Darwin thought of natural selection by analogy to how farmers select crops or livestock for breeding, which he called "artificial selection"; in his early manuscripts he referred to a *Nature*, which would do the selection. At the time, other mechanisms of evolution such as evolution by genetic drift were not yet explicitly formulated, and Darwin believed that selection was likely only part of the story: "I am convinced that Natural Selection has been the main but not exclusive means of modification." In a letter to Charles Lyell in September 1860, Darwin regretted the use of the term "Natural Selection," preferring the term "Natural Preservation."

For Darwin and his contemporaries, natural selection was in essence synonymous with evolution by natural selection. After the publication of *On the Origin of Species*, educated people generally accepted that evolution had occurred in some form. However, natural selection remained controversial as a mechanism, partly because it was perceived to be too weak to explain the range of observed characteristics of living organisms, and partly because even supporters of evolution balked at its "unguided" and non-progressive nature, a response that has been characterised as the single most significant impediment to the idea's acceptance.

However, some thinkers enthusiastically embraced natural selection; after reading Darwin, Herbert Spencer introduced the term *survival of the fittest*, which became a popular summary of the theory. The fifth edition of *On the Origin of Species* published in 1869 included Spencer's phrase as an alternative to natural selection, with credit given: "But the expression often used by Mr. Herbert Spencer of the Survival of the Fittest is more accurate, and is sometimes equally convenient." Although the phrase is still often used by non-biologists, modern biologists avoid it because it is tautological if "fittest" is read to mean "functionally superior" and is applied to individuals rather than considered as an averaged quantity over populations.

Modern Evolutionary Synthesis

Natural selection relies crucially on the idea of heredity, but developed before the basic concepts of genetics. Although the Moravian monk Gregor Mendel, the father of modern genetics, was a contemporary of Darwin's, his work would lie in obscurity until the early 20th century. Only after the 20th-century integration of Darwin's theory of evolution with a complex statistical appreciation of Gregor Mendel's "re-discovered" laws of inheritance did scientists generally come to accept natural selection.

The work of Ronald Fisher (who developed the required mathematical language and wrote *The Genetical Theory of Natural Selection* (1930)), J. B. S. Haldane (who introduced the concept of the "cost" of natural selection), Sewall Wright (who elucidated the nature of selection and adaptation), Theodosius Dobzhansky (who established the idea that mutation, by creating genetic diversity, supplied the raw material for natural selection: *Genetics and the Origin of Species* (1937)), William D. Hamilton (who conceived of kin selection), Ernst Mayr (who recognised the key importance of reproductive isolation for speciation: *Systematics and the Origin of Species* (1942)) and many others together formed the modern evolutionary synthesis. This synthesis cemented natural selection as the foundation of evolutionary theory, where it remains today.

Genetic Basis of Natural Selection

The idea of natural selection predates the understanding of genetics. We now have a much better idea of the biology underlying heritability, which is the basis of natural selection.

Genotype and Phenotype

Natural selection acts on an organism's phenotype, or physical characteristics. Phenotype is determined by an organism's genetic make-up (genotype) and the environment in which the organism lives. Often, natural selection acts on specific traits of an individual, and the terms phenotype and genotype are used narrowly to indicate these specific traits.

When different organisms in a population possess different versions of a gene for a certain trait, each of these versions is known as an allele. It is this genetic variation that underlies phenotypic traits. A typical example is that certain combinations of genes for eye colour in humans that, for instance, give rise to the phenotype of blue eyes. (On the other hand, when all the organisms in a population share the same allele for a particular trait, and this state is stable over time, the allele is said to be *fixed* in that population.)

Some traits are governed by only a single gene, but most traits are influenced by the interactions of many genes. A variation in one of the many genes that contributes to a trait may have only a small effect on the phenotype; together, these genes can produce a continuum of possible phenotypic values.

Directionality of Selection

When some component of a trait is heritable, selection will alter the frequencies of the different alleles, or variants of the gene that produces the variants of the trait. Selection can be divided into three classes, on the basis of its effect on allele frequencies.

Directional selection occurs when a certain allele has a greater fitness than others, resulting in an increase of its frequency. This process can continue until the allele is fixed and the entire population shares the fitter phenotype. It is directional selection that is illustrated in the antibiotic resistance example above.

Far more common is Stabilizing selection (which is commonly confused with *purifying selection*), which lowers the frequency of alleles that have a deleterious effect on the phenotype – that is, produce organisms of lower fitness. This process can continue until the allele is eliminated from the population. Purifying selection results in functional genetic features, such as protein-coding genes or regulatory sequences, being conserved over time due to selective pressure against deleterious variants.

A number of forms of balancing selection exist, which do not result in fixation, but maintain an allele at intermediate frequencies in a population. This can occur in diploid species (that is, those that have homologous pairs of chromosomes) when heterozygote individuals, who have different alleles on each chromosome at a single genetic locus, have a higher fitness than homozygote individuals that have two of the same alleles. This is called heterozygote advantage or over-dominance, of which the best-known example is the malarial resistance observed in heterozygous humans who carry only one copy of the gene for sickle-cell anaemia. Maintenance of allelic variation can also occur through disruptive or diversifying selection, which favours genotypes that depart from the average in either direction (that is, the opposite of over-dominance), and can result in a bimodal distribution of trait values. Finally, balancing selection can occur through frequency-dependent selection, where the fitness of one particular phenotype depends on the distribution of other phenotypes in the population. The principles of game theory have been applied to understand the fitness distributions in these situations, particularly in the study of kin selection and the evolution of reciprocal altruism.

Selection and Genetic Variation

A portion of all genetic variation is functionally neutral in that it produces no phenotypic effect or significant difference in fitness; the hypothesis that this variation accounts for a large fraction of observed genetic diversity is known as the neutral theory of molecular evolution and was originated by Motoo Kimura. When genetic variation does not result in differences in fitness, selection cannot *directly* affect the frequency of such variation. As a result, the genetic variation at those sites will be higher than at sites where variation does influence fitness. However, after a period with no new mutation, the genetic variation at these sites will be eliminated due to genetic drift.

Mutation–selection Balance

Natural selection results in the reduction of genetic variation through the elimination of maladapted individuals and consequently of the mutations that caused the maladaptation. At the same time, new mutations occur, resulting in a Mutation–selection balance. The exact outcome of the two processes depends both on the rate at which new mutations occur and on the strength of the natural selection, which is a function of how unfavourable the mutation proves to be. Consequently, changes in the mutation rate or the selection pressure will result in a different Mutation–selection balance.

Genetic Linkage

Genetic linkage occurs when the loci of two alleles are *linked*, or in close proximity to each other on the chromosome. During the formation of gametes, recombination of the genetic material results in reshuffling of the alleles. However, the chance that such a reshuffle occurs between two alleles depends on the distance between those alleles; the closer the alleles are to each other, the less likely it is that such a reshuffle will occur. Consequently, when selection targets one allele, this automatically results in selection of the other allele as well; through this mechanism, selection can have a strong influence on patterns of variation in the genome.

Selective sweeps occur when an allele becomes more common in a population as a result of positive selection. As the prevalence of one allele increases, linked alleles can also become more common, whether they are neutral or even slightly deleterious. This is called *genetic hitchhiking*. A strong selective sweep results in a region of the genome where the positively selected haplotype (the allele and its neighbours) are in essence the only ones that exist in the population.

Whether a selective sweep has occurred or not can be investigated by measuring linkage disequilibrium, or whether a given haplotype is overrepresented in the population. Normally, genetic recombination results in a reshuffling of the different alleles within a haplotype, and none of the haplotypes will dominate the population. However, during a selective sweep, selection for a specific allele will also result in selection of neighbouring alleles. Therefore, the presence of a block of strong linkage disequilibrium might indicate that there has been a 'recent' selective sweep near the centre of the block, and this can be used to identify sites recently under selection.

Background selection is the opposite of a selective sweep. If a specific site experiences strong and persistent purifying selection, linked variation will tend to be weeded out along with it, producing a region in the genome of low overall variability. Because background selection is a result of deleterious new mutations, which can occur randomly in any haplotype, it does not produce clear blocks of linkage disequilibrium, although with low recombination it can still lead to slightly negative linkage disequilibrium overall.

Competition

In the context of natural selection, competition is an interaction between organisms or species in which the fitness of one is lowered by the presence of another. Limited supply of at least one resource (such as food, water, and territory) used by both can be a factor. Competition; both within and between species is an important topic in ecology, especially community ecology. Competition is one of many interacting biotic and abiotic factors that affect community structure. Competition among members of the same species is known as intraspecific competition, while competition between individuals of different species is known as interspecific competition. Competition is not always straightforward, and can occur in both a direct and indirect fashion.

According to the competitive exclusion principle, species less suited to compete for resources should either adapt or die out, although competitive exclusion is rarely found in natural ecosystems. According to evolutionary theory, this competition within and between species for resources plays a very relevant role in natural selection, however, competition may play less of a role than expansion among larger clades, this is termed the 'Room to Roam' hypothesis.

In evolutionary contexts, competition is related to the concept of r/K selection theory, which relates to the selection of traits which promote success in particular environments. The theory originates from work on island biogeography by the ecologists Robert H. MacArthur and Edward O. Wilson.

In r/K selection theory, selective pressures are hypothesised to drive evolution in one of two stereotyped directions: r- or K-selection. These terms, r and K, are derived from standard ecological algebra, as illustrated in the simple Verhulst equation of population dynamics:

$$\frac{dN}{dt} = rN\left(1 - \frac{N}{K}\right)$$

where r is the growth rate of the population (N), and K is the carrying capacity of its local environmental setting. Typically, r-selected species exploit empty niches, and produce many offspring, each of whom has a relatively low probability of surviving to adulthood. In contrast, K-selected species are strong competitors in crowded niches, and invest more heavily in much fewer offspring, each of whom has a relatively high probability of surviving to adulthood.

Impact of the Idea

Darwin's ideas, along with those of Adam Smith and Karl Marx, had a profound influence on 19th century thought. Perhaps the most radical claim of the theory of evolution through natural selection is that "...elaborately constructed forms, so different from each other, and dependent on each other in so complex a manner, ..." evolved from the simplest forms of life by a few simple principles. This claim inspired some of Darwin's most ardent supporters—and provoked the most profound opposition. The radicalism of natural selection, according to Stephen Jay Gould, lay in its power to "dethrone some of the deepest and most traditional comforts of Western thought." In particular, it challenged long-standing beliefs in such concepts as a special and exalted place for humans in the natural world and a benevolent creator whose intentions were reflected in nature's order and design.

In the words of the philosopher Daniel Dennett, "Darwin's dangerous idea" of evolution by natural selection is a "universal acid," which cannot be kept restricted to any vessel or container, as it soon leaks out, working its way into ever-wider surroundings. Thus, in the last decades, the concept of natural selection has spread from evolutionary biology into virtually all disciplines, including evolutionary computation, quantum Darwinism, evolutionary economics, evolutionary epistemology, evolutionary psychology, and cosmological natural selection. This unlimited applicability has been called universal Darwinism.

Emergence of Natural Selection

How life originated from inorganic matter (abiogensis) remains an unresolved problem in biology. One prominent hypothesis is that life first appeared in the form of short self-replicating RNA polymers. (RNA world hypothesis) On this view, "life" may have come into existence when RNA chains first experienced the basic conditions, as conceived by Charles Darwin, for natural selection to operate. These conditions are: heritability, variation of type, and competition for limited resources. Fitness of an early RNA replicator (its per capita rate of increase) would likely have been a function of adaptive capacities that were intrinsic (i.e., determined by the nucleotide sequence) and the availability of resources. The three primary adaptive capacities could logically have been: (1) the capacity to replicate with moderate fidelity (giving rise to both heritability and variation of type), (2) the capacity to avoid decay, and (3) the capacity to acquire and process resources. These capacities would have been determined initially by the folded configurations (including those configurations with ribozyme activity) of the RNA replicators that, in turn, would have been encoded in their individual nucleotide sequences. Competitive success among different RNA replicators would have depended on the relative values of their adaptive capacities.

Cell and Molecular Biology

In 1881, Wilhelm Roux, a founder of modern embryology, published *Der Kampf der Theile im Organismus* (*The Struggle of Parts in the Organism*) in which he suggested that the development of an organism results from a Darwinian competition between the parts of the embryo, occurring at all levels, from molecules to organs. In recent years, a modern version of this theory has been proposed by Jean-Jacques Kupiec. According to this cellular Darwinism, Stochasticity at the molecular level generates diversity in cell types whereas cell interactions impose a characteristic order on the developing embryo.

Social and Psychological Theory

The social implications of the theory of evolution by natural selection also became the source of continuing controversy. Friedrich Engels, a German political philosopher and co-originator of the ideology of communism, wrote in 1872 that "Darwin did not know what a bitter satire he wrote on mankind, and especially on his countrymen, when he showed that free competition, the struggle for existence, which the economists celebrate as the highest historical achievement, is the normal state of the *animal kingdom*." Interpretation of natural selection, by Herbert Spencer and Francis Galton as necessarily "progressive," leading to increasing "advances" in intelligence and civilisation, was used as a justification for colonialism and policies of eugenics, as well as to support broader sociopolitical positions now described as social Darwinism. Konrad Lorenz won the Nobel

Prize in Physiology or Medicine in 1973 for his analysis of animal behaviour in terms of the role of natural selection (particularly group selection). However, in Germany in 1940, in writings that he subsequently disowned, he used the theory as a justification for policies of the Nazi state. He wrote "... selection for toughness, heroism, and social utility...must be accomplished by some human institution, if mankind, in default of selective factors, is not to be ruined by domestication-induced degeneracy. The racial idea as the basis of our state has already accomplished much in this respect." Others have developed ideas that human societies and culture evolve by mechanisms analogous to those that apply to evolution of species.

More recently, work among anthropologists and psychologists has led to the development of sociobiology and later of evolutionary psychology, a field that attempts to explain features of human psychology in terms of adaptation to the ancestral environment. The most prominent example of evolutionary psychology, notably advanced in the early work of Noam Chomsky and later by Steven Pinker, is the hypothesis that the human brain has adapted to acquire the grammatical rules of natural language. Other aspects of human behaviour and social structures, from specific cultural norms such as incest avoidance to broader patterns such as gender roles, have been hypothesised to have similar origins as adaptations to the early environment in which modern humans evolved. By analogy to the action of natural selection on genes, the concept of memes—"units of cultural transmission," or culture's equivalents of genes undergoing selection and recombination—has arisen, first described in this form by Richard Dawkins in 1976 and subsequently expanded upon by philosophers such as Daniel Dennett as explanations for complex cultural activities, including human consciousness.

Information and Systems Theory

In 1922, Alfred J. Lotka proposed that natural selection might be understood as a physical principle that could be described in terms of the use of energy by a system, a concept that was later developed by Howard Odum as the maximum power principle whereby evolutionary systems with selective advantage maximise the rate of useful energy transformation. Such concepts are sometimes relevant in the study of applied thermodynamics.

The principles of natural selection have inspired a variety of computational techniques, such as "soft" artificial life, that simulate selective processes and can be highly efficient in 'adapting' entities to an environment defined by a specified fitness function. For example, a class of heuristic optimisation algorithms known as genetic algorithms, pioneered by John Henry Holland in the 1970s and expanded upon by David E. Goldberg, identify optimal solutions by simulated reproduction and mutation of a population of solutions defined by an initial probability distribution. Such algorithms are particularly useful when applied to problems whose energy landscape is very rough or has many local minima.

References

- Rieger R. Michaelis A., Green M. M. (1976): Glossary of genetics and cytogenetics: Classical and molecular. Springer-Verlag, Heidelberg - New York, ISBN 3-540-07668-9; ISBN 0-387-07668-9.

- Cavali-Sforza L. L., Bodmer W. F. (1999): The genetics of human populations. Dover Publications, Inc., Mineola, New York, ISBN 0-486-40693-8.

- Mayr E. (1970): Populations, species, and evolution – An abridgment of Animal species and evolution. The

Belknap Press of Harvard University Press, Cambridge, Massachusetts and London, England, ISBN 0-674-69013-3.

- Robinson, Richard, ed. (2003). "Population Bottleneck". Genetics. 3. New York: Macmillan Reference USA. ISBN 0-02-865609-1. LCCN 2002003560. OCLC 614996575. Retrieved 2015-12-14.

- Eyre-Walker, Adam; Keightley, Peter D. (August 2007). "The distribution of fitness effects of new mutations" (PDF). Nature Reviews Genetics. London: Nature Publishing Group. 8 (8): 610–618. doi:10.1038/nrg2146. ISSN 1471-0056. PMID 17637733. Retrieved 2015-10-02.

- Orr, H. Allen (April 2003). "The Distribution of Fitness Effects Among Beneficial Mutations". Genetics. Bethesda, MD: Genetics Society of America. 163 (4): 1519–1526. ISSN 0016-6731. PMC 1462510. PMID 12702694. Retrieved 2015-10-07.

- Hogan, C. Michael (October 12, 2010). "Mutation". In Monosson, Emily. Encyclopedia of Earth. Washington, D.C.: Environmental Information Coalition, National Council for Science and the Environment. OCLC 72808636. Retrieved 2015-10-08.

- Darwin, Charles (September 28, 1860). "Darwin, C. R. to Lyell, Charles". Darwin Correspondence Project. Cambridge, UK: Cambridge University Library. Letter 2931. Retrieved 2015-08-01.

- Haldane, J. B. S. (December 1957). "The Cost of Natural Selection" (PDF). Journal of Genetics. 55 (3): 511–524. doi:10.1007/BF02984069. ISSN 0022-1333. Retrieved 2015-08-02.

- Wright, Sewall (1932). "The roles of mutation, inbreeding, crossbreeding and selection in evolution". Proceedings of the VI International Congress of Genetics. 1: 356–366. Retrieved 2015-08-02.

- Gould, Stephen Jay (June 12, 1997). "Darwinian Fundamentalism". The New York Review of Books. New York: Rea S. Hederman. 44 (10). ISSN 0028-7504. Retrieved 2015-08-03.

- Cornuet, Jean Marie; Luikart, Gordon (December 1996). "Description and Power Analysis of Two Tests for Detecting Recent Population Bottlenecks from Allele Frequency Data". Genetics. Bethesda, MD: Genetics Society of America. 144 (4): 2001–2014. ISSN 0016-6731.

4

Human Genetics: A Comprehensive Study

With the breakthroughs in genetics that have been a direct result of technological advancement, it comes as no surprise that the euchromatic areas of the human genome have been completely mapped by The Human Genome Project (HGP). This has enabled scientists around the globe to better understand genetic disorders and to also pursue improved gene therapy. This chapter focuses on the leaps and bounds achieved by human genetics and equips the reader with detailed information about the cutting-edge developments in human genetics.

Human Genetics

Human Genetics is the study of inheritance as it occurs in human beings. Human genetics encompasses a variety of overlapping fields including: classical genetics, cytogenetics, molecular genetics, biochemical genetics, genomics, population genetics, developmental genetics, clinical genetics, and genetic counseling.

Genes can be the common factor of the qualities of most human-inherited traits. Study of human genetics can be useful as it can answer questions about human nature, understand the diseases and development of effective disease treatment, and understand genetics of human life. This article describes only basic features of human genetics; for the genetics of disorders please see: Medical genetics.

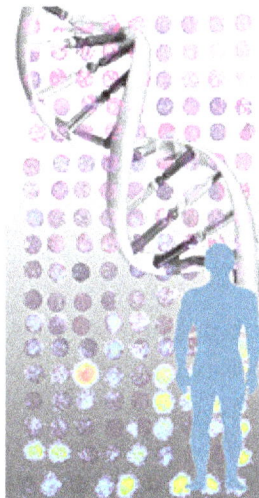

A small piece of human DNA

Genetic Differences and Inheritance Patterns

Inheritance of traits for humans are based upon Gregor Mendel's model of inheritance. Mendel deduced that inheritance depends upon discrete units of inheritance, called factors or genes.

Autosomal Dominant Inheritance

Autosomal traits are associated with a single gene on an autosome (non-sex chromosome)—they are called "dominant" because a single copy—inherited from either parent—is enough to cause this trait to appear. This often means that one of the parents must also have the same trait, unless it has arisen due to an unlikely new mutation. Examples of autosomal dominant traits and disorders are Huntington's disease and achondroplasia.

Autosomal Recessive Inheritance

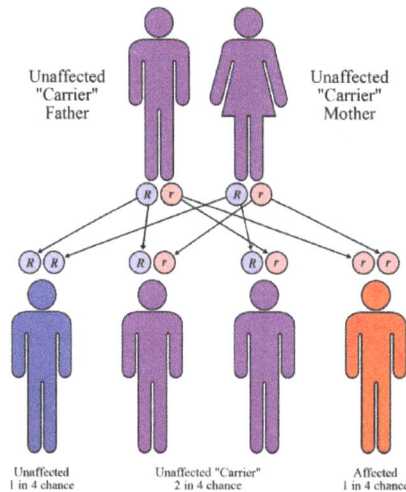

Autosomal recessive inheritance, a 25% chance

Autosomal recessive traits is one pattern of inheritance for a trait, disease, or disorder to be passed on through families. For a recessive trait or disease to be displayed two copies of the trait or disorder needs to be presented. The trait or gene will be located on a non-sex chromosome. Because it takes two copies of a trait to display a trait, many people can unknowingly be carriers of a disease. From an evolutionary perspective, a recessive disease or trait can remain hidden for several generations before displaying the phenotype. Examples of autosomal recessive disorders are albinism, cystic fibrosis.

X-linked and Y-linked Inheritance

X-linked genes are found on the sex X chromosome. X-linked genes just like autosomal genes have both dominant and recessive types. Recessive X-linked disorders are rarely seen in females and usually only affect males. This is because males inherit their X chromosome and all X-linked genes will be inherited from the maternal side. Fathers only pass on their Y chromosome to their sons, so no X-linked traits will be inherited from father to son. Men cannot be carriers for recessive X linked traits, as they only have one X chromosome, so any X linked trait inherited from the mother will show up.

Females express X-linked disorders when they are homozygous for the disorder and become carriers when they are heterozygous. X-linked dominant inheritance will show the same phenotype as a heterozygote and homozygote. Just like X-linked inheritance, there will be a lack of male-to-male inheritance, which makes it distinguishable from autosomal traits. One example of an X-linked trait is Coffin–Lowry syndrome, which is caused by a mutation in ribosomal protein gene. This mutation results in skeletal, craniofacial abnormalities, mental retardation, and short stature.

X chromosomes in females undergo a process known as X inactivation. X inactivation is when one of the two X chromosomes in females is almost completely inactivated. It is important that this process occurs otherwise a woman would produce twice the amount of normal X chromosome proteins. The mechanism for X inactivation will occur during the embryonic stage. For people with disorders like trisomy X, where the genotype has three X chromosomes, X-inactivation will inactivate all X chromosomes until there is only one X chromosome active. Males with Klinefelter syndrome, who have an extra X chromosome, will also undergo X inactivation to have only one completely active X chromosome.

Y-linked inheritance occurs when a gene, trait, or disorder is transferred through the Y chromosome. Since Y chromosomes can only be found in males, Y linked traits are only passed on from father to son. The testis determining factor, which is located on the Y chromosome, determines the maleness of individuals. Besides the maleness inherited in the Y-chromosome there are no other found Y-linked characteristics.

Pedigrees

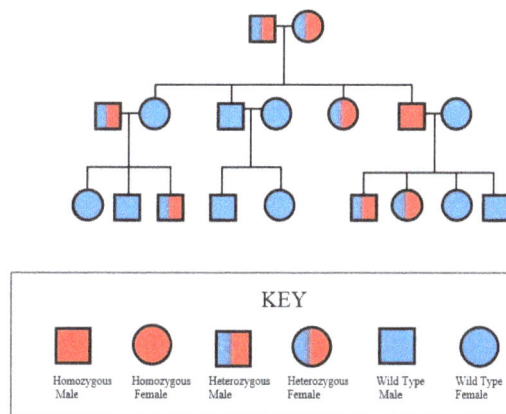

An example of a family pedigree displaying an autosomal recessive trait

A pedigree is a diagram showing the ancestral relationships and transmission of genetic traits over several generations in a family. Square symbols are almost always used to represent males, whilst circles are used for females. Pedigrees are used to help detect many different genetic diseases. A pedigree can also be used to help determine the chances for a parent to produce an offspring with a specific trait.

Four different traits can be identified by pedigree chart analysis: autosomal dominant, autosomal recessive, x-linked, or y-linked. Partial penetrance can be shown and calculated form pedigrees. Penetrance is the percentage expressed frequency with which individuals of a given genotype manifest at least some degree of a specific mutant phenotype associated with a trait.

Inbreeding, or mating between closely related organisms, can clearly be seen on pedigree charts.

Pedigree charts of royal families often have a high degree of inbreeding, because it was customary and preferable for royalty to marry another member of royalty. Genetic counselors commonly use pedigrees to help couples determine if the parents will be able to produce healthy children.

Karyotype

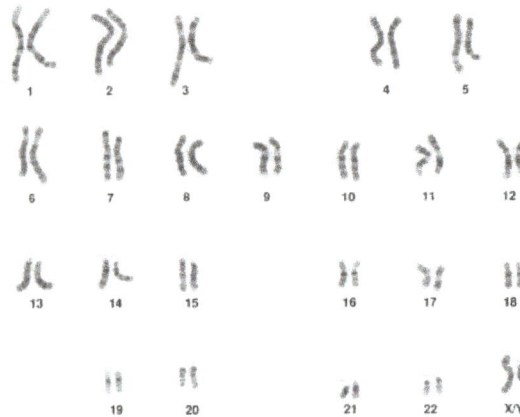

A karyotype of a human male, showing 46 chromosomes including XY sex chromosomes.

A karyotype is a very useful tool in cytogenetics. A karyotype is picture of all the chromosomes in the metaphase stage arranged according to length and centromere position. A karyotype can also be useful in clinical genetics, due to its ability to diagnose genetic disorders. On a normal karyotype, aneuploidy can be detected by clearly being able to observe any missing or extra chromosomes.

Giemsa banding, g-banding, of the karyotype can be used to detect deletions, insertions, duplications, inversions, and translocations. G-banding will stain the chromosomes with light and dark bands unique to each chromosome. A FISH, fluorescent in situ hybridization, can be used to observe deletions, insertions, and translocations. FISH uses fluorescent probes to bind to specific sequences of the chromosomes that will cause the chromosomes to fluoresce a unique color.

Genomics

Genomics refers to the field of genetics concerned with structural and functional studies of the genome. A genome is all the DNA contained within an organism or a cell including nuclear and mitochondrial DNA. The human genome is the total collection of genes in a human being contained in the human chromosome, composed of over three billion nucleotides. In April 2003, the Human Genome Project was able to sequence all the DNA in the human genome, and to discover that the human genome was composed of around 20,000 protein coding genes.

Medical Genetics

Medical genetics' is the branch of medicine that involves the diagnosis and management of hereditary disorders. Medical genetics is the application of genetics to medical care. It overlaps human genetics, for example, research on the causes and inheritance of genetic disorders would be considered within both human genetics and medical genetics, while the diagnosis, management, and counseling of individuals with genetic disorders would be considered part of medical genetics.

Population Genetics

Population genetics is the branch of evolutionary biology responsible for investigating processes that cause changes in allele and genotype frequencies in populations based upon Mendelian inheritance. Four different forces can influence the frequencies: natural selection, mutation, gene flow (migration), and genetic drift. A population can be defined as a group of interbreeding individuals and their offspring. For human genetics the populations will consist only of the human species. The Hardy-Weinberg principle is a widely used principle to determine allelic and genotype frequencies.

Mitochondrial DNA

In addition to nuclear DNA, humans (like almost all eukaryotes) have mitochondrial DNA. Mitochondria, the "power houses" of a cell, have their own DNA. Mitochondria are inherited from one's mother, and its DNA is frequently used to trace maternal lines of descent. Mitochondrial DNA is only 16kb in length and encodes for 62 genes.

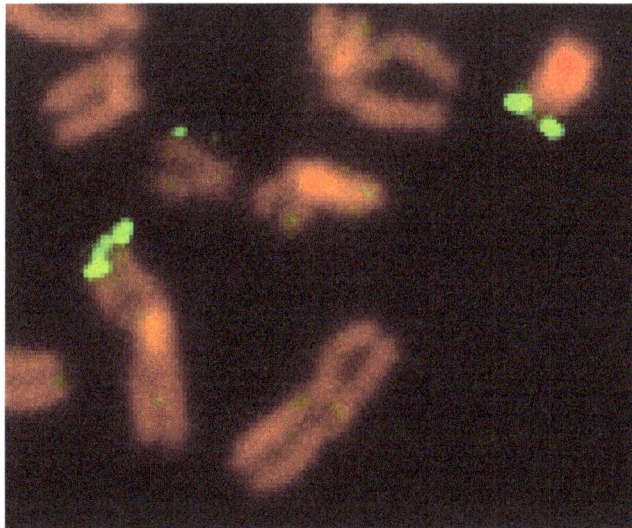

XY Chromosomes

Genes and Sex

The XY sex-determination system is the sex-determination system found in humans, most other mammals, some insects (*Drosophila*), and some plants (*Ginkgo*). In this system, the sex of an individual is determined by a pair of sex chromosomes (gonosomes). Females have two of the same kind of sex chromosome (XX), and are called the homogametic sex. Males have two distinct sex chromosomes (XY), and are called the heterogametic sex.

X-linked Traits

Sex linkage is the phenotypic expression of an allele related to the chromosomal sex of the individual. This mode of inheritance is in contrast to the inheritance of traits on autosomal chromosomes, where both sexes have the same probability of inheritance. Since humans have many more genes on the X than the Y, there are many more X-linked traits than Y-linked traits. However, females carry two or more copies of the X chromosome, resulting in a potentially toxic dose of X-linked genes.

To correct this imbalance, mammalian females have evolved a unique mechanism of dosage compensation. In particular, by way of the process called X-chromosome inactivation (XCI), female mammals transcriptionally silence one of their two Xs in a complex and highly coordinated manner.

X-link dominant	X-link recessive	References
Alport syndrome	Absence of blood in urine	
Coffin–Lowry syndrome	No cranial malformations	
Colour vision	Colour blindness	
Normal clotting factor	Haemophilia A & B	
Strong muscle tissue	Duchenne Muscular Dystrophy	
fragile X syndrome	Normal X chromosome	
Aicardi syndrome	Absence of brain defects	
Absence of autoimmunity	IPEX syndrome	
Xg blood type	Absence of antigen	
Production of GAGs	Hunter syndrome	
Normal muscle strength	Becker's Muscular Dystrophy	
Unaffected body	Fabry's disease	
No progressive blindness	Choroideremia	
No kidney damage	Dent's disease	
Rett syndrome	No microcephaly	
Production of HGPRT	Lesch–Nyhan syndrome	
High levels of copper	Menkes disease	
Normal immune levels	Wiskott–Aldrich syndrome	
Focal dermal hypoplasia	Normal pigmented skin	
Normal pigment in eyes	Ocular albinism	
Vitamin D resistant rickets	Absorption of Vitamin D	
Synesthesia	Non colour perception	

Human Traits with Simple Inheritance Patterns

Dominant	Recessive	References
Low heart rate	High heart rate	
Widow's peak	straight hair line	
ocular hypertelorism	Hypotelorism	
Normal digestive muscle	POLIP syndrome	
Facial dimples *	No facial dimples	
Able to taste PTC	Unable to taste PTC	

Unattached (free) earlobe	Attached earlobe	
Clockwise hair direction (left to right)	Counter-Clockwise hair direction (right to left)	
Cleft chin	smooth chin	
No progressive nerve damage	Friedreich's ataxia	
Ability to roll tongue (Able to hold tongue in a U shape)	No ability to roll tongue	
extra finger or toe	Normal five fingers and toes	
Straight Thumb	Hitchhiker's Thumb	
Freckles	No freckles	
Wet-type earwax	Dry-type earwax	
Normal flat palm	Cenani Lenz syndactylism	
shortness in fingers	Normal finger length	
Webbed fingers	Normal separated fingers	
Roman nose	No prominent bridge	
Marfan's syndrome	Normal body proportions	
Huntington's disease	No nerve damage	
Normal mucus lining	Cystic fibrosis	
Photic sneeze reflex	No ACHOO reflex	
Forged chin	Receding chin	
White Forelock	Dark Forelock	
Ligamentous angustus	Ligamentous Laxity	
Ability to eat sugar	Galactosemia	
Total leukonychia and Bart pumphrey syndrome	partial leukonychia	
Absence of fish-like body odour	Trimethylaminuria	
Primary Hyperhidrosis	little sweating in hands	
Lactose persistence *	Lactose intolerance *	
Prominent chin (V-shaped)	less prominent chin (U-shaped)	
Acne prone	Clear complexion	
Normal height	Cartilage–hair hypoplasia	

Handicapping Conditions

Genetic Chromosomal

Effect	Source	References
Down syndrome	Additional 21st chromosome	
Cri du chat syndrome	Partial deletion of a chromosome in the B Group	
Klinefelter syndrome	One or more extra sex chromosome(s)	
Turner syndrome	Rearrangement of one or both X chromosomes, deletion of part of the second X chromosome, presence of part of a Y chromosome	

Human Evolutionary Genetics

Human evolutionary genetics studies how one human genome differs from another human genome, the evolutionary past that gave rise to it, and its current effects. Differences between genomes have anthropological, medical and forensic implications and applications. Genetic data can provide important insight into human evolution.

Origin of Apes

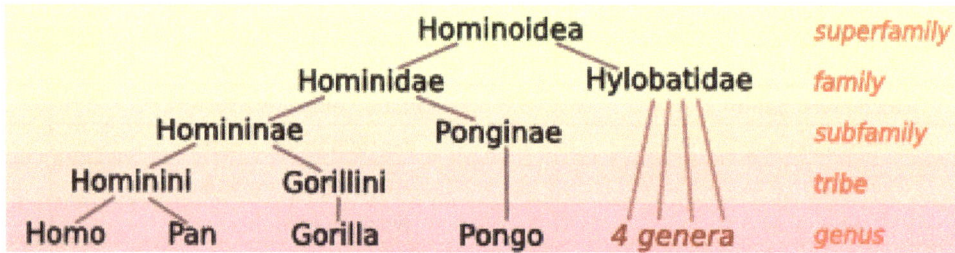

The taxonomic relationships of hominoids.

Biologists classify humans, along with only a few other species, as great apes (species in the family Hominidae). The living Hominidae include two distinct species of chimpanzee (the bonobo, *Pan paniscus*, and the common chimpanzee, *Pan troglodytes*), two species of gorilla (the western gorilla, *Gorilla gorilla*, and the eastern gorilla, *Gorilla graueri*), and two species of orangutan (the Bornean orangutan, *Pongo pygmaeus*, and the Sumatran orangutan, *Pongo abelii*). The great apes with the family Hylobatidae of gibbons form the superfamily Hominoidea of apes.

Apes, in turn, belong to the primate order (>400 species), along with the Old World monkeys, the New World monkeys, and others. Data from both mitochondrial DNA (mtDNA) and nuclear DNA (nDNA) indicate that primates belong to the group of Euarchontoglires, together with Rodentia, Lagomorpha, Dermoptera, and Scandentia. This is further supported by Alu-like short interspersed nuclear elements (SINEs) which have been found only in members of the Euarchontoglires.

Cladistics

A phylogenetic tree is usually derived from DNA or protein sequences from populations. Often, mitochondrial DNA or Y chromosome sequences are used to study ancient human demographics. These single-locus sources of DNA do not recombine and are almost always inherited from a single parent, with only one known exception in mtDNA. Individuals from closer geographic regions generally tend to be more similar than individuals from regions farther away. Distance on a phylogenetic tree can be used approximately to indicate:

1. Genetic distance. The genetic difference between humans and chimps is less than 2%, or twenty times larger than the variation among modern humans.

2. Temporal remoteness of the most recent common ancestor. The mitochondrial most recent common ancestor of modern humans lived roughly 200,000 years ago, latest common ancestors of humans and chimps between four and seven million years ago.

Speciation of Humans and the African Apes

The separation of humans from their closest relatives, the non-human apes (chimpanzees and gorillas), has been studied extensively for more than a century. Five major questions have been addressed:

- Which apes are our closest ancestors?

- When did the separations occur?

- What was the effective population size of the common ancestor before the split?

- Are there traces of population structure (subpopulations) preceding the speciation or partial admixture succeeding it?

- What were the specific events (including fusion of chromosomes 2a and 2b) prior to and subsequent to the separation?

General Observations

As discussed before, different parts of the genome show different sequence divergence between different hominoids. It has also been shown that the sequence divergence between DNA from humans and chimpanzees varies greatly. For example, the sequence divergence varies between 0% to 2.66% between non-coding, non-repetitive genomic regions of humans and chimpanzees. Additionally gene trees, generated by comparative analysis of DNA segments, do not always fit the species tree. Summing up:

- The sequence divergence varies significantly between humans, chimpanzees and gorillas.

- For most DNA sequences, humans and chimpanzees appear to be most closely related, but some point to a human-gorilla or chimpanzee-gorilla clade.

- The human genome has been sequenced, as well as the chimpanzee genome. Humans have 23 pairs of chromosomes, while chimpanzees, gorillas, and orangutans have 24. Human chromosome 2 is a fusion of two chromosomes 2a and 2b that remained separate in the other primates.

Divergence Times

The divergence time of humans from other apes is of great interest. One of the first molecular studies, published in 1967 measured immunological distances (IDs) between different primates. Basically the study measured the strength of immunological response that an antigen from one species (human albumin) induces in the immune system of another species (human, chimpanzee, gorilla and Old World monkeys). Closely related species should have similar antigens and therefore weaker immunological response to each other's antigens. The immunological response of a species to its own antigens (e.g. human to human) was set to be 1.

The ID between humans and gorillas was determined to be 1.09, that between humans and chimpanzees was determined as 1.14. However the distance to six different Old World monkeys was on average 2.46, indicating that the African apes are more closely related to humans than to monkeys.

The authors consider the divergence time between Old World monkeys and hominoids to be 30 million years ago (MYA), based on fossil data, and the immunological distance was considered to grow at a constant rate. They concluded that divergence time of humans and the African apes to be roughly ~5 MYA. That was a surprising result. Most scientists at that time thought that humans and great apes diverged much earlier (>15 MYA).

The gorilla was, in ID terms, closer to human than to chimpanzees; however, the difference was so slight that the trichotomy could not be resolved with certainty. Later studies based on molecular genetics were able to resolve the trichotomy: chimpanzees are phylogenetically closer to humans than to gorillas. However, some divergence times estimated later (using much more sophisticated methods in molecular genetics) do not substantially differ from the very first estimate in 1967, but a recent paper puts it at 11-14 MYA.

Divergence Times and Ancestral Effective Population Size

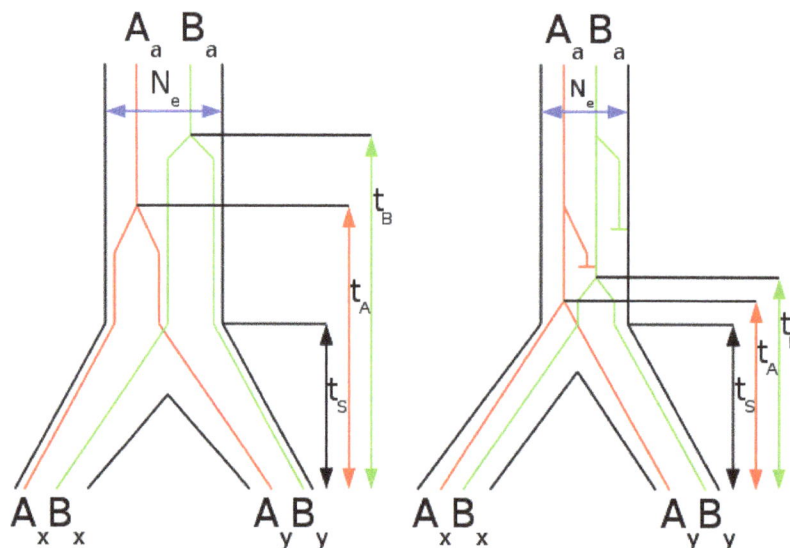

The sequences of the DNA segments diverge earlier than the species. A large effective population size in the ancestral population (left) preserves different variants of the DNA segments (=alleles) for a longer period of time. Therefore, on average, the gene divergence times (tA for DNA segment A; tB for DNA segment B) will deviate more from the time the species diverge (tS) compared to a small ancestral effective population size (right).

Current methods to determine divergence times use DNA sequence alignments and molecular clocks. Usually the molecular clock is calibrated assuming that the orangutan split from the African apes (including humans) 12-16 MYA. Some studies also include some old world monkeys and set the divergence time of them from hominoids to 25-30 MYA. Both calibration points are based on very little fossil data and have been criticized.

If these dates are revised, the divergence times estimated from molecular data will change as well. However, the relative divergence times are unlikely to change. Even if we can't tell absolute divergence times exactly, we can be pretty sure that the divergence time between chimpanzees and humans is about sixfold shorter than between chimpanzees (or humans) and monkeys.

One study (Takahata et al., 1995) used 15 DNA sequences from different regions of the genome from human and chimpanzee and 7 DNA sequences from human, chimpanzee and gorilla. They determined that chimpanzees are more closely related to humans than gorillas. Using various statistical methods, they estimated the divergence time human-chimp to be 4.7 MYA and the divergence time between gorillas and humans (and chimps) to be 7.2 MYA.

Additionally they estimated the effective population size of the common ancestor of humans and chimpanzees to be ~100,000. This was somewhat surprising since the present day effective population size of humans is estimated to be only ~10,000. If true that means that the human lineage would have experienced an immense decrease of its effective population size (and thus genetic diversity) in its evolution.

Another study (Chen & Li, 2001) sequenced 53 non-repetitive, intergenic DNA segments from a human, a chimpanzee, a gorilla, and orangutan. When the DNA sequences were concatenated to a single long sequence, the generated neighbor-joining tree supported the *Homo-Pan* clade with 100% bootstrap (that is that humans and chimpanzees are the closest related species of the four). When three species are fairly closely related to each other (like human, chimpanzee and gorilla), the trees obtained from DNA sequence data may not be congruent with the tree that represents the speciation (species tree).

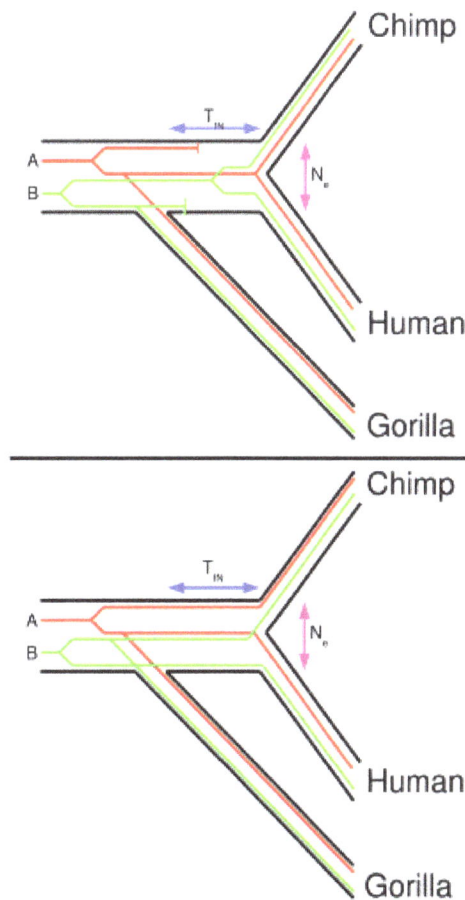

A and B are two different loci. In the upper figure they fit to the species tree. The DNA that is present in today's gorillas diverged earlier from the DNA that is present in today's humans and chimps. Thus both loci should be more similar between human and chimp than between gorilla and chimp or gorilla and human. In the lower graph, locus A has a more recent common ancestor in human and gorilla compared to the chimp sequence. Whereas chimp and gorilla have a more recent common ancestor for locus B. Here the gene trees are incongruent to the species tree.

The shorter internodal time span (T$_{IN}$) the more common are incongruent gene trees. The effective population size (N$_e$) of the internodal population determines how long genetic lineages are preserved in the population. A higher effective population size causes more incongruent gene trees. Therefore, if the internodal time span is known, the ancestral effective population size of the common ancestor of humans and chimpanzees can be calculated.

When each segment was analyzed individually, 31 supported the *Homo-Pan* clade, 10 supported the *Homo-Gorilla* clade, and 12 supported the *Pan-Gorilla* clade. Using the molecular clock the authors estimated that gorillas split up first 6.2-8.4 MYA and chimpanzees and humans split up 1.6-2.2 million years later (internodal time span) 4.6-6.2 MYA. The internodal time span is useful to estimate the ancestral effective population size of the common ancestor of humans and chimpanzees.

A parsimonious analysis revealed that 24 loci supported the *Homo-Pan* clade, 7 supported the *Homo-Gorilla* clade, 2 supported the *Pan-Gorilla* clade and 20 gave no resolution. Additionally they took 35 protein coding loci from databases. Of these 12 supported the *Homo-Pan* clade, 3 the *Homo-Gorilla* clade, 4 the *Pan-Gorilla* clade and 16 gave no resolution. Therefore, only ~70% of the 52 loci that gave a resolution (33 intergenic, 19 protein coding) support the 'correct' species tree. From the fraction of loci which did not support the species tree and the internodal time span they estimated previously, the effective population of the common ancestor of humans and chimpanzees was estimated to be ~52 000 to 96 000. This value is not as high as that from the first study (Takahata), but still much higher than present day effective population size of humans.

A third study (Yang, 2002) used the same dataset that Chen and Li used but estimated the ancestral effective population of 'only' ~12,000 to 21,000, using a different statistical method.

Genetic Differences Between Humans and Other Great Apes

The alignable sequences within genomes of humans and chimpanzees differ by about 35 million single-nucleotide substitutions. Additionally about 3% of the complete genomes differ by deletions, insertions and duplications.

Since mutation rate is relatively constant, roughly one half of these changes occurred in the human lineage. Only a very tiny fraction of those fixed differences gave rise to the different phenotypes of humans and chimpanzees and finding those is a great challenge. The vast majority of the differences are neutral and do not affect the phenotype.

Molecular evolution may act in different ways, through protein evolution, gene loss, differential gene regulation and RNA evolution. All are thought to have played some part in human evolution.

Gene Loss

Many different mutations can inactivate a gene, but few will change its function in a specific way. Inactivation mutations will therefore be readily available for selection to act on. Gene loss could thus be a common mechanism of evolutionary adaptation (the "less-is-more" hypothesis).

80 genes were lost in the human lineage after separation from the last common ancestor with the

chimpanzee. 36 of those were for olfactory receptors. Genes involved in chemoreception and immune response are overrepresented. Another study estimated that 86 genes had been lost.

Hair Keratin Gene Krthap1

A gene for type I hair keratin was lost in the human lineage. Keratins are a major component of hairs. Humans still have nine functional type I hair keratin genes, but the loss of that particular gene may have caused the thinning of human body hair. The gene loss occurred relatively recently in human evolution—less than 240,000 years ago.

Myosin Gene MYH16

Stedman *et al.* (2004) stated that the loss of the sarcomeric myosin gene MYH16 in the human lineage led to smaller masticatory muscles. They estimated that the mutation that led to the inactivation (a two base pair deletion) occurred 2.4 million years ago, predating the appearance of *Homo ergaster/erectus* in Africa. The period that followed was marked by a strong increase in cranial capacity, promoting speculation that the loss of the gene may have removed an evolutionary constraint on brain size in the genus *Homo*.

Another estimate for the loss of the MYH16 gene is 5.3 million years ago, long before *Homo* appeared.

Other

- CASPASE12, a cysteinyl aspartate proteinase. The loss of this gene is speculated to have reduced the lethality of bacterial infection in humans.

Gene Addition

Segmental duplications (SDs or LCRs) have had roles in creating new primate genes and shaping human genetic variation.

Human-specific DNA Insertions

When the human genome was compared to the genomes of five comparison primate species, including the chimpanzee, gorilla, orangutan, gibbon, and macaque, it was found that there are approximately 20,000 human-specific insertions believed to be regulatory. While most insertions appear to be fitness neutral, a small amount have been identified in positively selected genes showing associations to neural phenotypes and some relating to dental and sensory perception-related phenotypes. These findings hint at the seemingly important role of human-specific insertions in the recent evolution of humans.

Selection Pressures

Human accelerated regions are areas of the genome that differ between humans and chimpanzees to a greater extent than can be explained by genetic drift over the time since the two species shared a common ancestor. These regions show signs of being subject to natural selection, leading to the evolution of distinctly human traits. Two examples are HAR1F, which is believed to be related to brain development and HAR2 (a.k.a. HACNS1) that may have played a role in the development of the opposable thumb.

It has also been hypothesized that much of the difference between humans and chimpanzees is attributable to the regulation of gene expression rather than differences in the genes themselves. Analyses of conserved non-coding sequences, which often contain functional and thus positively selected regulatory regions, address this possibility.

Sequence Divergence Between Humans and Apes

When the draft sequence of the common chimpanzee (*Pan troglodytes*) genome was published in the summer 2005, 2400 million bases (of ~3160 million bases) were sequenced and assembled well enough to be compared to the human genome. 1.23% of this sequenced differed by single-base substitutions. Of this, 1.06% or less was thought to represent fixed differences between the species, with the rest being variant sites in humans or chimpanzees. Another type of difference, called indels (insertions/deletions) accounted for many fewer differences (15% as many), but contributed ~1.5% of unique sequence to each genome, since each insertion or deletion can involve anywhere from one base to millions of bases.

A companion paper examined segmental duplications in the two genomes, whose insertion and deletion into the genome account for much of the indel sequence. They found that a total of 2.7% of euchromatic sequence had been differentially duplicated in one or the other lineage.

Percentage sequence divergence between humans and other hominids			
Locus	Human-Chimp	Human-Gorilla	Human-Orangutan
Alu elements	2	-	-
Non-coding (Chr. Y)	1.68 ± 0.19	2.33 ± 0.2	5.63 ± 0.35
Pseudogenes (autosomal)	1.64 ± 0.10	1.87 ± 0.11	-
Pseudogenes (Chr. X)	1.47 ± 0.17	-	-
Noncoding (autosomal)	1.24 ± 0.07	1.62 ± 0.08	3.08 ± 0.11
Genes (K_s)	1.11	1.48	2.98
Introns	0.93 ± 0.08	1.23 ± 0.09	-
Xq13.3	0.92 ± 0.10	1.42 ± 0.12	3.00 ± 0.18
Subtotal for X chromosome	1.16 ± 0.07	1.47 ± 0.08	-
Genes (K_a)	0.8	0.93	1.96

The sequence divergence has generally the following pattern: Human-Chimp < Human-Gorilla << Human-Orangutan, highlighting the close kinship between humans and the African apes. Alu elements diverge quickly due to their high frequency of CpG dinucleotides which mutate roughly 10 times more often than the average nucleotide in the genome. The mutation rate is higher in the male germ line, therefore the divergence in the Y chromosome—which is inherited solely from the father—is higher than in autosomes. The X chromosome is inherited twice as often through the female germ line as through the male germ line and therefore shows slightly lower sequence divergence. The sequence divergence of the Xq13.3 region is surprisingly low between humans and chimpanzees.

Mutations altering the amino acid sequence of proteins (K$_a$) are the least common. In fact ~29% of all orthologous proteins are identical between human and chimpanzee. The typical protein differs by only two amino acids. The measures of sequence divergence shown in the table only take the substitutional differences, for example from an A (adenine) to a G (guanine), into account. DNA sequences may however also differ by insertions and deletions (indels) of bases. These are usually stripped from the alignments before the calculation of sequence divergence is performed.

Genetic Differences between Modern Humans and Neanderthals

An international group of scientists completed a draft sequence of the Neanderthal genome in May 2010. The results indicate some breeding between modern humans (*Homo sapiens*) and Neanderthals (*Homo neanderthalensis*), as the genomes of non-African humans have 1-4% more in common with Neanderthals than do the genomes of subsaharan Africans. Neanderthals and most modern humans share a lactose-intolerant variant of the lactase gene that encodes an enzyme that is unable to break down lactose in milk after weaning. Modern humans and Neanderthals also share the FOXP2 gene variant associated with brain development and with speech in modern humans, indicating that Neanderthals may have been able to speak. Chimps have two amino acid differences in FOXP2 compared with human and Neanderthal FOXP2.

Modern Humans

Molecular biologists starting with Wesley Brown on mtDNA and Allan Wilson on mtDNA have produced observations relevant to human evolution.

Age of the Common Ancestor

By estimating the rate at which mutations occur in mtDNA, the age of the common ancestral mtDNA type can be estimated: "the common ancestral mtDNA (type a) links mtDNA types that have diverged by an average of nearly 0.57%. Assuming a rate of 2%-4% per million years, this implies that the common ancestor of all surviving mtDNA types existed 140,000-290,000 years ago."

Most, but not all (fr. Multiregional origin of modern humans), scientists in the relevant fields consider this observation robust. This common direct ancestor in the line of mother to daughter (or mitochondrial most recent common ancestor (mtMRCA)) of all extant humans has become known as Mitochondrial Eve. (Mitochondria are inherited from the mother only.) The observation that the mtMRCA is the direct matrilineal ancestor of all living humans does not mean either that she was the first anatomically modern woman, nor that no other women lived concurrently with her, nor even excluding the existence of other women being ancestors of today's people. Other women would have lived at the same time and passed nuclear genes down to living humans, but their mitochondrial lineages were lost over time. This could be due to events such as producing only male children.

African Origin for Modern Humans

There is evidence that modern human mtDNA has an African origin: "We infer from the tree of minimum length... that Africa is a likely source of the human mitochondrial gene pool. This inference comes from the observation that one of the two primary branches leads exclusively to African mtDNAs... while the second primary branch also leads to African mtDNAs... By postulating that the common ancestral mtDNA... was African, we minimize the number of intercontinental migrations needed to account for the geographic distribution of mtDNA types."

The broad study of African genetic diversity headed by Sarah Tishkoff found the San people to express the greatest genetic diversity among the 113 distinct populations sampled, making them one of 14 "ancestral population clusters". The research also located the origin of modern human migration in south-western Africa, near the coastal border of Namibia and Angola.

Y Chromosome Lineages

The Y chromosome is much larger than mtDNA, and is relatively homogeneous; therefore it has taken much longer to find distinct lineages and to analyse them. Conversely, because the Y chromosome is so large by comparison, it holds more genetic information. Y chromosome studies show similar findings to those made with mtDNA. The estimate for the age of the ancestral Y chromosome for all extant Y chromosomes is given at about 70,000 years ago and is also placed in Africa; the individual who contributed this Y chromosomal heritage is sometimes referred to as Y chromosome Adam. The difference in dates between Y chromosome Adam and mitochondrial Eve is usually attributed to a higher extinction rate for Y chromosomes due to greater differential reproductive success between individual men, which means that a small number of very successful men may produce many children, while a larger number of less successful men will produce far fewer children.

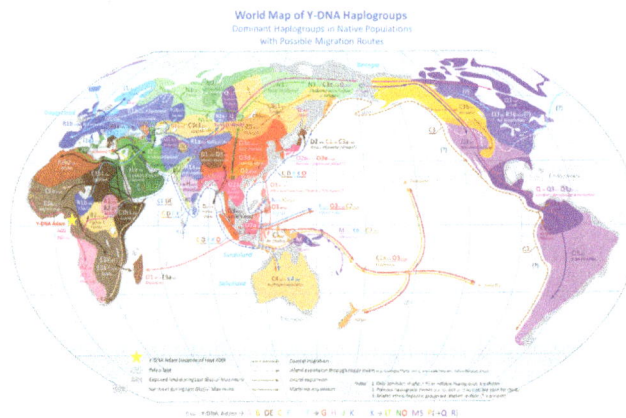

Dominant Y-chromosome haplogroups in pre-colonial world populations, with possible migrations routes according to the Coastal Migration Model.

At a recent conference at the American Society of Human Genetics, there was a study presented by Melissa Wilson Sayres from the University of California Berkeley which called into question the amount to which people had more female ancestors than male. While it didn't deny that men have historically produced at a higher rate than females, a polygamous mating system doesn't explain how the y-chromosome is so uniform. "Other people have suggested it's as few as four females for one male. We find that it's probably skewed, but it's more like four females to three males" said Sayres.

Ancestral Components

Modern DNA sequencing has identified various ancestral components in contemporary populations. A number of these genetic elements have West Eurasian origins. They include the following ancestral components, with their geographical hubs and main associated populations:

#	West Eurasian component	Geographical hub	Peak population	Notes
1	Ancestral North Indian	North India, Pakistan	North Indians, Pakistanis	Main West Eurasian component in the Indian subcontinent. Peaks among Indo-European-speaking caste populations in the northern areas, but also found at significant frequencies among some Dravidian-speaking caste groups. Associated with either the arrival of Indo-European speakers from West Asia or Central Asia between 3,000 and 4,000 years before present, or with the spread of agriculture and West Asian crops beginning around 8,000-9,000 ybp, or with migrations from West Asia in the pre-agricultural period. Contrasted with the indigenous Ancestral South Indian component, which peaks among the Onge Andamanese inhabiting the Andaman Islands.
2	Arabian	Arabian peninsula	Yemenis, Saudis, Qataris, Bedouins	Main West Eurasian component in the Gulf region. Most closely associated with local Arabic, Semitic-speaking populations. Also found at significant frequencies in parts of the Levant, Egypt and Libya.
3	Coptic	Nile Valley	Copts, Beja, Afro-Asiatic Ethiopians, Sudanese Arabs, Nubians	Main West Eurasian component in Northeast Africa. Roughly equivalent with the Ethio-Somali component. Peaks among Egyptian Copts in Sudan. Also found at high frequencies among other Afro-Asiatic (Hamito-Semitic) speakers in Ethiopia and Sudan, as well as among many Nubians. Associated with Ancient Egyptian ancestry, without the later Arabian influence present among modern Egyptians. Contrasted with the indigenous Nilo-Saharan component, which peaks among Nilo-Saharan- and Kordofanian-speaking populations inhabiting the southern part of the Nile Valley.
4	Ethio-Somali	Horn of Africa	Somalis, Afars, Amhara, Oromos, Tigrinya	Main West Eurasian component in the Horn. Roughly equivalent with the Coptic component. Associated with the arrival of Afro-Asiatic speakers in the region during antiquity. Peaks among Cushitic- and Ethiopian Semitic-speaking populations in the northern areas. Diverged from the Maghrebi component around 23,000 ybp, and from the Arabian component about 25,000 ybp. Contrasted with the indigenous Omotic component, which peaks among the Omotic-speaking Ari ironworkers inhabiting southern Ethiopia.
5	European	Europe	Europeans	Main West Eurasian component in Europe. Also found at significant frequencies in adjacent geographical areas outside of the continent, in Anatolia, the Caucasus, the Iranian plateau, and parts of the Levant.

#	West Eurasian component	Geographical hub	Peak population	Notes
6	Levantine	Near East, Caucasus	Druze, Lebanese, Cypriots, Syrians, Jordanians, Palestinians, Armenians, Georgians, Sephardic Jews, Ashkenazi Jews, Iranians, Turks, Sardinians, Adygei	Main West Eurasian component in the Near East and Caucasus. Peaks among Druze populations in the Levant. Found amongst local Afro-Asiatic, Indo-European, Caucasus and Turkish speakers alike. Diverged from the European component around 9,100-15,900 ybp, and from the Arabian component about 15,500-23,700 ypb. Also found at significant frequencies in Southern Europe as well as parts of the Arabian peninsula.
7	Maghrebi	Northwest Africa	Berbers, Maghrebis, Sahrawis, Tuareg	Main West Eurasian component in the Maghreb. Peaks among the Berber (non-Arabized) populations in the region. Diverged from the Ethio-Somali/Coptic, Arabian, Levantine and European components prior to the Holocene.

Human Genome

The human genome is the complete set of nucleic acid sequence for humans (*Homo sapiens*), encoded as DNA within the 23 chromosome pairs in cell nuclei and in a small DNA molecule found within individual mitochondria. Human genomes include both protein-coding DNA genes and noncoding DNA. Haploid human genomes, which are contained in germ cells (the egg and sperm gamete cells created in the meiosis phase of sexual reproduction before fertilization creates a zygote) consist of three billion DNA base pairs, while diploid genomes (found in somatic cells) have twice the DNA content. While there are significant differences among the genomes of human individuals (on the order of 0.1%), these are considerably smaller than the differences between humans and their closest living relatives, the chimpanzees (approximately 4%) and bonobos.

The Human Genome Project produced the first complete sequences of individual human genomes, with the first draft sequence and initial analysis being published on February 12, 2001. The human genome was the first of all vertebrates to be completely sequenced. As of 2012, thousands of human genomes have been completely sequenced, and many more have been mapped at lower levels of resolution. The resulting data are used worldwide in biomedical science, anthropology, forensics and other branches of science. There is a widely held expectation that genomic studies will lead to advances in the diagnosis and treatment of diseases, and to new insights in many fields of biology, including human evolution.

Although the sequence of the human genome has been (almost) completely determined by DNA sequencing, it is not yet fully understood. Most (though probably not all) genes have been identified by a combination of high throughput experimental and bioinformatics approaches, yet much work still needs to be done to further elucidate the biological functions of their protein and RNA products. Recent results suggest that most of the vast quantities of noncoding DNA within the genome have associated biochemical activities, including regulation of gene expression, organization of chromosome architecture, and signals controlling epigenetic inheritance.

There are an estimated 20,000-25,000 human protein-coding genes. The estimate of the number of human genes has been repeatedly revised down from initial predictions of 100,000 or more as genome sequence quality and gene finding methods have improved, and could continue to drop further. Protein-coding sequences account for only a very small fraction of the genome (approximately 1.5%), and the rest is associated with non-coding RNA molecules, regulatory DNA sequences, LINEs, SINEs, introns, and sequences for which as yet no function has been determined.

In June 2016, scientists formally announced HGP-Write, a plan to synthesize the human genome.

Molecular Organization and Gene Content

The total length of the human genome is over 3 billion base pairs. The genome is organized into 22 paired chromosomes, plus the X chromosome (one in males, two in females) and, in males only, one Y chromosome. These are all large linear DNA molecules contained within the cell nucleus. The genome also includes the mitochondrial DNA, a comparatively small circular molecule present in each mitochondrion. Basic information about these molecules and their gene content, based on a reference genome that does not represent the sequence of any specific individual, are provided in the following table. (Data source: Ensembl genome browser release 68, July 2012)

Chromo-some	Length (mm)	Base pairs	Variations	Con-firmed proteins	Putative proteins	Pseudo-genes	miRNA	rRNA	snR-NA	snoR-NA	Misc ncRNA	Links	Cen-tromere position (Mbp)	Cu-mu-lative (%)
1	85	249,250,621	4,401,091	2,012	31	1,130	134	66	221	145	106	EBI	125.0	7.9
2	83	243,199,373	4,607,702	1,203	50	948	115	40	161	117	93	EBI	93.3	16.2
3	67	198,022,430	3,894,345	1,040	25	719	99	29	138	87	77	EBI	91.0	23.0
4	65	191,154,276	3,673,892	718	39	698	92	24	120	56	71	EBI	50.4	29.6
5	62	180,915,260	3,436,667	849	24	676	83	25	106	61	68	EBI	48.4	35.8
6	58	171,115,067	3,360,890	1,002	39	731	81	26	111	73	67	EBI	61.0	41.6
7	54	159,138,663	3,045,992	866	34	803	90	24	90	76	70	EBI	59.9	47.1
8	50	146,364,022	2,890,692	659	39	568	80	28	86	52	42	EBI	45.6	52.0
9	48	141,213,431	2,581,827	785	15	714	69	19	66	51	55	EBI	49.0	56.3
10	46	135,534,747	2,609,802	745	18	500	64	32	87	56	56	EBI	40.2	60.9
11	46	135,006,516	2,607,254	1,258	48	775	63	24	74	76	53	EBI	53.7	65.4
12	45	133,851,895	2,482,194	1,003	47	582	72	27	106	62	69	EBI	35.8	70.0
13	39	115,169,878	1,814,242	318	8	323	42	16	45	34	36	EBI	17.9	73.4
14	36	107,349,540	1,712,799	601	50	472	92	10	65	97	46	EBI	17.6	76.4
15	35	102,531,392	1,577,346	562	43	473	78	13	63	136	39	EBI	19.0	79.3
16	31	90,354,753	1,747,136	805	65	429	52	32	53	58	34	EBI	36.6	82.0
17	28	81,195,210	1,491,841	1,158	44	300	61	15	80	71	46	EBI	24.0	84.8
18	27	78,077,248	1,448,602	268	20	59	32	13	51	36	25	EBI	17.2	87.4
19	20	59,128,983	1,171,356	1,399	26	181	110	13	29	31	15	EBI	26.5	89.3
20	21	63,025,520	1,206,753	533	13	213	57	15	46	37	34	EBI	27.5	91.4
21	16	48,129,895	787,784	225	8	150	16	5	21	19	8	EBI	13.2	92.6
22	17	51,304,566	745,778	431	21	308	31	5	23	23	23	EBI	14.7	93.8
X	53	155,270,560	2,174,952	815	23	780	128	22	85	64	52	EBI	60.6	99.1

Y	20	59,373,566	286,812	45	8	327	15	7	17	3	2	EBI	12.5	100.0
mtDNA	0.0054	16,569	929	13	0	0	0	2	0	0	22	EBI	N/A	100.0
total		3,095,693,981		19,313	738	12,859								

Table 1 (above) summarizes the physical organization and gene content of the human reference genome, with links to the original analysis, as published in the Ensembl database at the European Bioinformatics Institute (EBI) and Wellcome Trust Sanger Institute. Chromosome lengths were estimated by multiplying the number of base pairs by 0.34 nanometers, the distance between base pairs in the DNA double helix. The number of proteins is based on the number of initial precursor mRNA transcripts, and does not include products of alternative pre-mRNA splicing, or modifications to protein structure that occur after translation.

The number of variations is a summary of unique DNA sequence changes that have been identified within the sequences analyzed by Ensembl as of July, 2012; that number is expected to increase as further personal genomes are sequenced and examined. In addition to the gene content shown in this table, a large number of non-expressed functional sequences have been identified throughout the human genome. Links open windows to the reference chromosome sequence in the EBI genome browser. The table also describes prevalence of genes encoding structural RNAs in the genome.

MicroRNA, or miRNA, functions as a post-transcriptional regulator of gene expression. Ribosomal RNA, or rRNA, makes up the RNA portion of the ribosome and is critical in the synthesis of proteins. Small nuclear RNA, or snRNA, is found in the nucleus of the cell. Its primary function is in the processing of pre-mRNA molecules and also in the regulation of transcription factors. Small nucleolar RNA, or SnoRNA, primarily functions in guiding chemical modifications to other RNA molecules.

Completeness of the Human Genome Sequence

Although the human genome has been completely sequenced for all practical purposes, there are still hundreds of gaps in the sequence. A recent study noted more than 160 euchromatic gaps of which 50 gaps were closed. However, there are still numerous gaps in the heterochromatic parts of the genome which is much harder to sequence due to numerous repeats and other intractable sequence features.

Coding vs. Noncoding DNA

The content of the human genome is commonly divided into coding and noncoding DNA sequences. Coding DNA is defined as those sequences that can be transcribed into mRNA and translated into proteins during the human life cycle; these sequences occupy only a small fraction of the genome (<2%). Noncoding DNA is made up of all of those sequences (ca. 98% of the genome) that are not used to encode proteins.

Some noncoding DNA contains genes for RNA molecules with important biological functions (noncoding RNA, for example ribosomal RNA and transfer RNA). The exploration of the function and evolutionary origin of noncoding DNA is an important goal of contemporary genome research, including the ENCODE (Encyclopedia of DNA Elements) project, which aims to survey the entire human genome, using a variety of experimental tools whose results are indicative of molecular activity.

Because non-coding DNA greatly outnumbers coding DNA, the concept of the sequenced genome has become a more focused analytical concept than the classical concept of the DNA-coding gene.

Mutation Rate of Human Genome

Mutation rate of human genome is a very important factor in calculating evolutionary time points. Researchers calculated the number of genetic variations between human and apes. Dividing that number by age of fossil of most recent common ancestor of humans and ape, researchers calculated the mutation rate. Recent studies using next generation sequencing technologies concluded a slow mutation rate which doesn't add up with human migration pattern time points and suggesting a new evolutionary time scale. 100,000 year old human fossils found in Israel have served to compound this new found uncertainty of the human migration timeline.

Coding Sequences (Protein-coding Genes)

Protein-coding sequences represent the most widely studied and best understood component of the human genome. These sequences ultimately lead to the production of all human proteins, although several biological processes (e.g. DNA rearrangements and alternative pre-mRNA splicing) can lead to the production of many more unique proteins than the number of protein-coding genes.

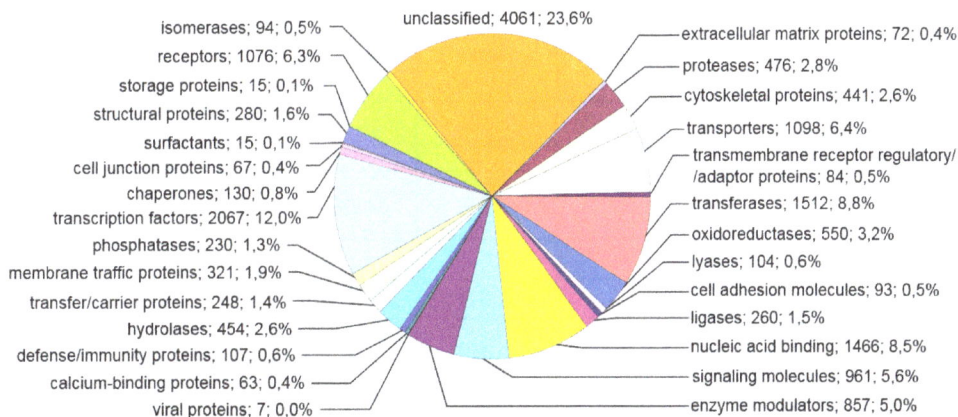

Human genes categorized by function of the transcribed proteins, given both as number of encoding genes and percentage of all genes.

The complete modular protein-coding capacity of the genome is contained within the exome, and consists of DNA sequences encoded by exons that can be translated into proteins. Because of its biological importance, and the fact that it constitutes less than 2% of the genome, sequencing of the exome was the first major milepost of the Human Genome Project.

Number of protein-coding genes. About 20,000 human proteins have been annotated in databases such as Uniprot. Historically, estimates for the number of protein genes have varied widely, ranging up to 2,000,000 in the late 1960s, but several researchers pointed out in the early 1970s that the estimated mutational load from deleterious mutations placed an upper limit of approximately 40,000 for the total number of functional loci (this includes protein-coding and functional non-coding genes).

The number of human protein-coding genes is not significantly larger than that of many less complex organisms, such as the roundworm and the fruit fly. This difference may result from the extensive use of alternative pre-mRNA splicing in humans, which provides the ability to build a very large number of modular proteins through the selective incorporation of exons.

Protein-coding capacity per chromosome. Protein-coding genes are distributed unevenly across the chromosomes, ranging from a few dozen to more than 2000, with an especially high gene density within chromosomes 19, 11, and 1 (Table 1). Each chromosome contains various gene-rich and gene-poor regions, which may be correlated with chromosome bands and GC-content. The significance of these nonrandom patterns of gene density is not well understood.

Size of protein-coding genes. The size of protein-coding genes within the human genome shows enormous variability (Table 2). For example, the gene for histone H1a (HIST1HIA) is relatively small and simple, lacking introns and encoding mRNA sequences of 781 nt and a 215 amino acid protein (648 nt open reading frame). Dystrophin (DMD) is the largest protein-coding gene in the human reference genome, spanning a total of 2.2 MB, while Titin (TTN) has the longest coding sequence (114,414 bp), the largest number of exons (363), and the longest single exon (17,106 bp). Over the whole genome, the median size of an exon is 122 bp (mean = 145 bp), the median number of exons is 7 (mean = 8.8), and the median coding sequence encodes 367 amino acids (mean = 447 amino acids; Table 21 in).

Protein	Chrom	Gene	Length	Exons	Exon length	Intron length	Alt splicing
Breast cancer type 2 susceptibility protein	13	BRCA2	83,736	27	11,386	72,350	yes
Cystic fibrosis transmembrane conductance regulator	7	CFTR	202,881	27	4,440	198,441	yes
Cytochrome b	MT	MTCYB	1,140	1	1,140	0	no
Dystrophin	X	DMD	2,220,381	79	10,500	2,209,881	yes
Glyceraldehyde-3-phosphate dehydrogenase	12	GAPDH	4,444	9	1,425	3,019	yes
Hemoglobin beta subunit	11	HBB	1,605	3	626	979	no
Histone H1A	6	HIST1H1A	781	1	781	0	no
Titin	2	TTN	281,434	364	104,301	177,133	yes

Table 2. Examples of human protein-coding genes. Chrom, chromosome. Alt splicing, alternative pre-mRNA splicing. (Data source: Ensembl genome browser release 68, July 2012)

Noncoding DNA (ncDNA)

Noncoding DNA is defined as all of the DNA sequences within a genome that are not found within protein-coding exons, and so are never represented within the amino acid sequence of expressed proteins. By this definition, more than 98% of the human genomes is composed of ncDNA.

Numerous classes of noncoding DNA have been identified, including genes for noncoding RNA (e.g. tRNA and rRNA), pseudogenes, introns, untranslated regions of mRNA, regulatory DNA sequences, repetitive DNA sequences, and sequences related to mobile genetic elements.

Numerous sequences that are included within genes are also defined as noncoding DNA. These include genes for noncoding RNA (e.g. tRNA, rRNA), and untranslated components of protein-coding genes (e.g. introns, and 5' and 3' untranslated regions of mRNA).

Protein-coding sequences (specifically, coding exons) constitute less than 1.5% of the human genome. In addition, about 26% of the human genome is introns. Aside from genes (exons and introns) and known regulatory sequences (8–20%), the human genome contains regions of noncoding DNA. The exact amount of noncoding DNA that plays a role in cell physiology has been hotly debated. Recent analysis by the ENCODE project indicates that 80% of the entire human genome is either transcribed, binds to regulatory proteins, or is associated with some other biochemical activity.

It however remains controversial whether all of this biochemical activity contributes to cell physiology, or whether a substantial portion of this is the result transcriptional and biochemical noise, which must be actively filtered out by the organism. Excluding protein-coding sequences, introns, and regulatory regions, much of the non-coding DNA is composed of: Many DNA sequences that do not play a role in gene expression have important biological functions. Comparative genomics studies indicate that about 5% of the genome contains sequences of noncoding DNA that are highly conserved, sometimes on time-scales representing hundreds of millions of years, implying that these noncoding regions are under strong evolutionary pressure and positive selection.

Many of these sequences regulate the structure of chromosomes by limiting the regions of heterochromatin formation and regulating structural features of the chromosomes, such as the telomeres and centromeres. Other noncoding regions serve as origins of DNA replication. Finally several regions are transcribed into functional noncoding RNA that regulate the expression of protein-coding genes (for example), mRNA translation and stability, chromatin structure (including histone modifications, for example), DNA methylation (for example), DNA recombination (for example), and cross-regulate other noncoding RNAs (for example). It is also likely that many transcribed noncoding regions do not serve any role and that this transcription is the product of non-specific RNA Polymerase activity.

Pseudogenes

Pseudogenes are inactive copies of protein-coding genes, often generated by gene duplication, that have become nonfunctional through the accumulation of inactivating mutations. Table 1 shows that the number of pseudogenes in the human genome is on the order of 13,000, and in some chromosomes is nearly the same as the number of functional protein-coding genes. Gene duplication is a major mechanism through which new genetic material is generated during molecular evolution.

For example, the olfactory receptor gene family is one of the best-documented examples of pseudogenes in the human genome. More than 60 percent of the genes in this family are non-functional pseudogenes in humans. By comparison, only 20 percent of genes in the mouse olfactory receptor gene family are pseudogenes. Research suggests that this is a species-specific characteristic, as the most closely related primates all have proportionally fewer pseudogenes. This genetic discovery helps to explain the less acute sense of smell in humans relative to other mammals.

Genes for Noncoding RNA (ncRNA)

Noncoding RNA molecules play many essential roles in cells, especially in the many reactions of protein synthesis and RNA processing. Noncoding RNA include tRNA, ribosomal RNA, microRNA, snRNA and other non-coding RNA genes including about 60,000 long non coding RNAs (lncRNAs). It should be noted that while the number of reported lncRNA genes continues to rise and the exact number in the human genome is yet to be defined, many of them are argued to be non-functional.

Many ncRNAs are critical elements in gene regulation and expression. Noncoding RNA also contributes to epigenetics, transcription, RNA splicing, and the translational machinery. The role of RNA in genetic regulation and disease offers a new potential level of unexplored genomic complexity.

Introns and Untranslated Regions of mRNA

In addition to the ncRNA molecules that are encoded by discrete genes, the initial transcripts of protein coding genes usually contain extensive noncoding sequences, in the form of introns, 5'-untranslated regions (5'-UTR), and 3'-untranslated regions (3'-UTR). Within most protein-coding genes of the human genome, the length of intron sequences is 10- to 100-times the length of exon sequences (Table 2).

Regulatory DNA Sequences

The human genome has many different regulatory sequences which are crucial to controlling gene expression. Conservative estimates indicate that these sequences make up 8% of the genome, however extrapolations from the ENCODE project give that 20-40% of the genome is gene regulatory sequence. Some types of non-coding DNA are genetic "switches" that do not encode proteins, but do regulate when and where genes are expressed (called enhancers).

Regulatory sequences have been known since the late 1960s. The first identification of regulatory sequences in the human genome relied on recombinant DNA technology. Later with the advent of genomic sequencing, the identification of these sequences could be inferred by evolutionary conservation. The evolutionary branch between the primates and mouse, for example, occurred 70–90 million years ago. So computer comparisons of gene sequences that identify conserved non-coding sequences will be an indication of their importance in duties such as gene regulation.

Other genomes have been sequenced with the same intention of aiding conservation-guided methods, for exampled the pufferfish genome. However, regulatory sequences disappear and re-evolve during evolution at a high rate.

As of 2012, the efforts have shifted toward finding interactions between DNA and regulatory proteins by the technique ChIP-Seq, or gaps where the DNA is not packaged by histones (DNase hypersensitive sites), both of which tell where there are active regulatory sequences in the investigated cell type.

Repetitive DNA Sequences

Repetitive DNA sequences comprise approximately 50% of the human genome.

About 8% of the human genome consists of tandem DNA arrays or tandem repeats, low complexity repeat sequences that have multiple adjacent copies (e.g. "CAGCAGCAG..."). The tandem sequences may be of variable lengths, from two nucleotides to tens of nucleotides. These sequences are highly variable, even among closely related individuals, and so are used for genealogical DNA testing and forensic DNA analysis.

Repeated sequences of fewer than ten nucleotides (e.g. the dinucleotide repeat $(AC)_n$) are termed microsatellite sequences. Among the microsatellite sequences, trinucleotide repeats are of particular importance, as sometimes occur within coding regions of genes for proteins and may lead to genetic disorders. For example, Huntington's disease results from an expansion of the trinucleotide repeat $(CAG)_n$ within the *Huntingtin* gene on human chromosome 4. Telomeres (the ends of linear chromosomes) end with a microsatellite hexanucleotide repeat of the sequence $(TTAGGG)_n$.

Tandem repeats of longer sequences (arrays of repeated sequences 10–60 nucleotides long) are termed minisatellites.

Mobile Genetic Elements (Transposons) and their Relics

Transposable genetic elements, DNA sequences that can replicate and insert copies of themselves at other locations within a host genome, are an abundant component in the human genome. The most abundant transposon lineage, *Alu*, has about 50,000 active copies, and can be inserted into intragenic and intergenic regions. One other lineage, LINE-1, has about 100 active copies per genome (the number varies between people). Together with non-functional relics of old transposons, they account for over half of total human DNA. Sometimes called "jumping genes", transposons have played a major role in sculpting the human genome. Some of these sequences represent endogenous retroviruses, DNA copies of viral sequences that have become permanently integrated into the genome and are now passed on to succeeding generations.

Mobile elements within the human genome can be classified into LTR retrotransposons (8.3% of total genome), SINEs (13.1% of total genome) including Alu elements, LINEs (20.4% of total genome), SVAs and Class II DNA transposons (2.9% of total genome).

Genomic Variation in Humans

Human Reference Genome

With the exception of identical twins, all humans show significant variation in genomic DNA sequences. The Human Reference Genome (HRG) is used as a standard sequence reference.

There are several important points concerning the Human Reference Genome--

- The HRG is a haploid sequence. Each chromosome is represented once.

- The HRG is a composite sequence, and does not correspond to any actual human individual.

- The HRG is periodically updated to correct errors and ambiguities.

- The HRG in no way represents an "ideal" or "perfect" human individual. It is simply a standardized representation or model that is used for comparative purposes.

Measuring Human Genetic Variation

Most studies of human genetic variation have focused on single-nucleotide polymorphisms (SNPs), which are substitutions in individual bases along a chromosome. Most analyses estimate that SNPs occur 1 in 1000 base pairs, on average, in the euchromatic human genome, although they do not occur at a uniform density. Thus follows the popular statement that "we are all, regardless of race, genetically 99.9% the same", although this would be somewhat qualified by most geneticists. For example, a much larger fraction of the genome is now thought to be involved in copy number variation. A large-scale collaborative effort to catalog SNP variations in the human genome is being undertaken by the International HapMap Project.

The genomic loci and length of certain types of small repetitive sequences are highly variable from person to person, which is the basis of DNA fingerprinting and DNA paternity testing technologies. The heterochromatic portions of the human genome, which total several hundred million base pairs, are also thought to be quite variable within the human population (they are so repetitive and so long that they cannot be accurately sequenced with current technology). These regions contain few genes, and it is unclear whether any significant phenotypic effect results from typical variation in repeats or heterochromatin.

Most gross genomic mutations in gamete germ cells probably result in inviable embryos; however, a number of human diseases are related to large-scale genomic abnormalities. Down syndrome, Turner Syndrome, and a number of other diseases result from nondisjunction of entire chromosomes. Cancer cells frequently have aneuploidy of chromosomes and chromosome arms, although a cause and effect relationship between aneuploidy and cancer has not been established.

Mapping Human Genomic Variation

Whereas a genome sequence lists the order of every DNA base in a genome, a genome map identifies the landmarks. A genome map is less detailed than a genome sequence and aids in navigating around the genome.

An example of a variation map is the HapMap being developed by the International HapMap Project. The HapMap is a haplotype map of the human genome, "which will describe the common patterns of human DNA sequence variation." It catalogs the patterns of small-scale variations in the genome that involve single DNA letters, or bases.

Researchers published the first sequence-based map of large-scale structural variation across the human genome in the journal *Nature* in May 2008. Large-scale structural variations are differences in the genome among people that range from a few thousand to a few million DNA bases; some are gains or losses of stretches of genome sequence and others appear as re-arrangements of stretches of sequence. These variations include differences in the number of copies individuals have of a particular gene, deletions, translocations and inversions.

SNP Frequency Across the Human Genome

Single-nucleotide polymorphisms (SNPs) do not occur homogeneously across the human genome. In fact, there is enormous diversity in SNP frequency between genes, reflecting different selective pressures on each gene as well as different mutation and recombination rates across the genome.

However, studies on SNPs are biased towards coding regions, the data generated from them are unlikely to reflect the overall distribution of SNPs throughout the genome. Therefore, the SNP Consortium protocol was designed to identify SNPs with no bias towards coding regions and the Consortium's 100,000 SNPs generally reflect sequence diversity across the human chromosomes. The SNP Consortium aims to expand the number of SNPs identified across the genome to 300 000 by the end of the first quarter of 2001.

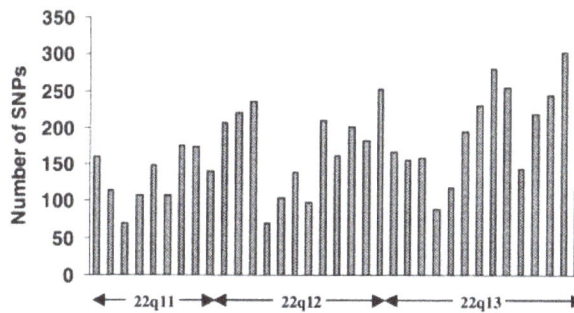

TSC SNP distribution along the long arm of chromosome 22 (from http://snp.cshl.org/). Each column represents a 1 Mb interval; the approximate cytogenetic position is given on the x-axis. Clear peaks and troughs of SNP density can be seen, possibly reflecting different rates of mutation, recombination and selection.

Changes in non-coding sequence and synonymous changes in coding sequence are generally more common than non-synonymous changes, reflecting greater selective pressure reducing diversity at positions dictating amino acid identity. Transitional changes are more common than transversions, with CpG dinucleotides showing the highest mutation rate, presumably due to deamination.

Personal Genomes

A personal genome sequence is a (nearly) complete sequence of the chemical base pairs that make up the DNA of a single person. Because medical treatments have different effects on different people due to genetic variations such as single-nucleotide polymorphisms (SNPs), the analysis of personal genomes may lead to personalized medical treatment based on individual genotypes.

The first personal genome sequence to be determined was that of Craig Venter in 2007. Personal genomes had not been sequenced in the public Human Genome Project to protect the identity of volunteers who provided DNA samples. That sequence was derived from the DNA of several volunteers from a diverse population. However, early in the Venter-led Celera Genomics genome sequencing effort the decision was made to switch from sequencing a composite sample to using DNA from a single individual, later revealed to have been Venter himself. Thus the Celera human genome sequence released in 2000 was largely that of one man. Subsequent replacement of the early composite-derived data and determination of the diploid sequence, representing both sets of chromosomes, rather than a haploid sequence originally reported, allowed the release of the first personal genome. In April 2008, that of James Watson was also completed. Since then hundreds of personal genome sequences have been released, including those of Desmond Tutu, and of a Paleo-Eskimo. In November 2013, a Spanish family made their personal genomics data obtained by direct-to-consumer genetic testing with 23andMe publicly available under a Creative Commons public domain license. This is believed to be the first such public genomics dataset for a whole family.

The sequencing of individual genomes further unveiled levels of genetic complexity that had not been appreciated before. Personal genomics helped reveal the significant level of diversity in the human genome attributed not only to SNPs but structural variations as well. However, the application of such knowledge to the treatment of disease and in the medical field is only in its very beginnings. Exome sequencing has become increasingly popular as a tool to aid in diagnosis of genetic disease because the exome contributes only 1% of the genomic sequence but accounts for roughly 85% of mutations that contribute significantly to disease.

Human Genetic Disorders

Most aspects of human biology involve both genetic (inherited) and non-genetic (environmental) factors. Some inherited variation influences aspects of our biology that are not medical in nature (height, eye color, ability to taste or smell certain compounds, etc.). Moreover, some genetic disorders only cause disease in combination with the appropriate environmental factors (such as diet). With these caveats, genetic disorders may be described as clinically defined diseases caused by genomic DNA sequence variation. In the most straightforward cases, the disorder can be associated with variation in a single gene. For example, cystic fibrosis is caused by mutations in the CFTR gene, and is the most common recessive disorder in caucasian populations with over 1,300 different mutations known.

Disease-causing mutations in specific genes are usually severe in terms of gene function, and are fortunately rare, thus genetic disorders are similarly individually rare. However, since there are many genes that can vary to cause genetic disorders, in aggregate they constitute a significant component of known medical conditions, especially in pediatric medicine. Molecularly characterized genetic disorders are those for which the underlying causal gene has been identified, currently there are approximately 2,200 such disorders annotated in the OMIM database.

Studies of genetic disorders are often performed by means of family-based studies. In some instances population based approaches are employed, particularly in the case of so-called founder populations such as those in Finland, French-Canada, Utah, Sardinia, etc. Diagnosis and treatment of genetic disorders are usually performed by a geneticist-physician trained in clinical/medical genetics. The results of the Human Genome Project are likely to provide increased availability of genetic testing for gene-related disorders, and eventually improved treatment. Parents can be screened for hereditary conditions and counselled on the consequences, the probability it will be inherited, and how to avoid or ameliorate it in their offspring.

As noted above, there are many different kinds of DNA sequence variation, ranging from complete extra or missing chromosomes down to single nucleotide changes. It is generally presumed that much naturally occurring genetic variation in human populations is phenotypically neutral, i.e. has little or no detectable effect on the physiology of the individual (although there may be fractional differences in fitness defined over evolutionary time frames). Genetic disorders can be caused by any or all known types of sequence variation. To molecularly characterize a new genetic disorder, it is necessary to establish a causal link between a particular genomic sequence variant and the clinical disease under investigation. Such studies constitute the realm of human molecular genetics.

With the advent of the Human Genome and International HapMap Project, it has become feasible to explore subtle genetic influences on many common disease conditions such as diabetes, asthma, migraine, schizophrenia, etc. Although some causal links have been made between genomic sequence variants in particular genes and some of these diseases, often with much publicity in the general media, these are usually not considered to be genetic disorders *per se* as their causes are complex, involving many different genetic and environmental factors. Thus there may be disagreement in particular cases whether a specific medical condition should be termed a genetic disorder. The categorized table below provides the prevalence as well as the genes or chromosomes associated with some human genetic disorders.

Disorder	Prevalence	Chromosome or gene involved
Chromosomal conditions		
Down syndrome	1:600	Chromosome 21
Klinefelter syndrome	1:500–1000 males	Additional X chromosome
Turner syndrome	1:2000 females	Loss of X chromosome
Sickle cell anemia	1 in 50 births in parts of Africa; rarer elsewhere	β-globin (on chromosome 11)
Cancers		
Breast/Ovarian cancer (susceptibility)	~5% of cases of these cancer types	BRCA1, BRCA2
FAP (hereditary nonpolyposis coli)	1:3500	APC
Lynch syndrome	5–10% of all cases of bowel cancer	MLH1, MSH2, MSH6, PMS2
Neurological conditions		
Huntington disease	1:20000	Huntingtin
Alzheimer disease - early onset	1:2500	PS1, PS2, APP
Other conditions		
Cystic fibrosis	1:2500	CFTR
Duchenne muscular dystrophy	1:3500 boys	Dystrophin

Evolution

Comparative genomics studies of mammalian genomes suggest that approximately 5% of the human genome has been conserved by evolution since the divergence of extant lineages approximately 200 million years ago, containing the vast majority of genes. The published chimpanzee genome differs from that of the human genome by 1.23% in direct sequence comparisons. Around 20% of this figure is accounted for by variation within each species, leaving only ~1.06% consistent sequence divergence between humans and chimps at shared genes. This nucleotide by nucleotide difference is dwarfed, however, by the portion of each genome that is not shared, including around 6% of functional genes that are unique to either humans or chimps.

In other words, the considerable observable differences between humans and chimps may be due as much or more to genome level variation in the number, function and expression of genes rather than DNA sequence changes in shared genes. Indeed, even within humans, there has been found to be a previously unappreciated amount of copy number variation (CNV) which can make up as much as 5 – 15% of the human genome. In other words, between humans, there could be +/-

500,000,000 base pairs of DNA, some being active genes, others inactivated, or active at different levels. The full significance of this finding remains to be seen. On average, a typical human protein-coding gene differs from its chimpanzee ortholog by only two amino acid substitutions; nearly one third of human genes have exactly the same protein translation as their chimpanzee orthologs. A major difference between the two genomes is human chromosome 2, which is equivalent to a fusion product of chimpanzee chromosomes 12 and 13. (later renamed to chromosomes 2A and 2B, respectively).

Humans have undergone an extraordinary loss of olfactory receptor genes during our recent evolution, which explains our relatively crude sense of smell compared to most other mammals. Evolutionary evidence suggests that the emergence of color vision in humans and several other primate species has diminished the need for the sense of smell.

Mitochondrial DNA

The human mitochondrial DNA is of tremendous interest to geneticists, since it undoubtedly plays a role in mitochondrial disease. It also sheds light on human evolution; for example, analysis of variation in the human mitochondrial genome has led to the postulation of a recent common ancestor for all humans on the maternal line of descent.

Due to the lack of a system for checking for copying errors, mitochondrial DNA (mtDNA) has a more rapid rate of variation than nuclear DNA. This 20-fold higher mutation rate allows mtDNA to be used for more accurate tracing of maternal ancestry. Studies of mtDNA in populations have allowed ancient migration paths to be traced, such as the migration of Native Americans from Siberia or Polynesians from southeastern Asia. It has also been used to show that there is no trace of Neanderthal DNA in the European gene mixture inherited through purely maternal lineage. Due to the restrictive all or none manner of mtDNA inheritance, this result (no trace of Neanderthal mtDNA) would be likely unless there were a large percentage of Neanderthal ancestry, or there was strong positive selection for that mtDNA (for example, going back 5 generations, only 1 of your 32 ancestors contributed to your mtDNA, so if one of these 32 was pure Neanderthal you would expect that ~3% of your autosomal DNA would be of Neanderthal origin, yet you would have a ~97% chance to have no trace of Neanderthal mtDNA).

Epigenome

Epigenetics describes a variety of features of the human genome that transcend its primary DNA sequence, such as chromatin packaging, histone modifications and DNA methylation, and which are important in regulating gene expression, genome replication and other cellular processes. Epigenetic markers strengthen and weaken transcription of certain genes but do not affect the actual sequence of DNA nucleotides. DNA methylation is a major form of epigenetic control over gene expression and one of the most highly studied topics in epigenetics. During development, the human DNA methylation profile experiences dramatic changes. In early germ line cells, the genome has very low methylation levels. These low levels generally describe active genes. As development progresses, parental imprinting tags lead to increased methylation activity.

Epigenetic patterns can be identified between tissues within an individual as well as between individuals themselves. Identical genes that have differences only in their epigenetic state are called epialleles. Epialleles can be placed into three categories: those directly determined by an individual's genotype, those influenced by genotype, and those entirely independent of genotype. The epigenome is also influenced significantly by environmental factors. Diet, toxins, and hormones impact the epigenetic state. Studies in dietary manipulation have demonstrated that methyl-deficient diets are associated with hypomethylation of the epigenome. Such studies establish epigenetics as an important interface between the environment and the genome.

Genetic Disorder

A genetic disorder is a genetic problem caused by one or more abnormalities in the genome, especially a condition that is present from birth (congenital). Most genetic disorders are quite rare and affect one person in every several thousands or millions.

Genetic disorders may be hereditary, passed down from the parents' genes. In other genetic disorders, defects may be caused by new mutations or changes to the DNA. In such cases, the defect will only be passed down if it occurs in the germ line. The same disease, such as some forms of cancer, may be caused by an inherited genetic condition in some people, by new mutations in other people, and mainly by environmental causes in other people. Whether, when and to what extent a person with the genetic defect or abnormality will actually suffer from the disease is almost always affected by the environmental factors and events in the person's development.

Some types of recessive gene disorders confer an advantage in certain environments when only one copy of the gene is present.

Single-gene

Prevalence of some single-gene disorders	
Disorder prevalence (approximate)	
Autosomal dominant	
Familial hypercholesterolemia	1 in 500
Polycystic kidney disease	1 in 1250
Neurofibromatosis type I	1 in 2,500
Hereditary spherocytosis	1 in 5,000
Marfan syndrome	1 in 4,000
Huntington's disease	1 in 15,000
Autosomal recessive	
Sickle cell anaemia	1 in 625
Cystic fibrosis	1 in 2,000
Tay-Sachs disease	1 in 3,000
Phenylketonuria	1 in 12,000
Mucopolysaccharidoses	1 in 25,000
Lysosomal acid lipase deficiency	1 in 40,000
Glycogen storage diseases	1 in 50,000

Galactosemia	1 in 57,000
X-linked	
Duchenne muscular dystrophy	1 in 7,000
Hemophilia	1 in 10,000
Values are for liveborn infants	

A single-gene disorder is the result of a single mutated gene. Over 4000 human diseases are caused by single-gene defects. Single-gene disorders can be passed on to subsequent generations in several ways. Genomic imprinting and uniparental disomy, however, may affect inheritance patterns. The divisions between recessive and dominant types are not "hard and fast", although the divisions between autosomal and X-linked types are (since the latter types are distinguished purely based on the chromosomal location of the gene). For example, achondroplasia is typically considered a dominant disorder, but children with two genes for achondroplasia have a severe skeletal disorder of which achondroplasics could be viewed as carriers. Sickle-cell anemia is also considered a recessive condition, but heterozygous carriers have increased resistance to malaria in early childhood, which could be described as a related dominant condition. When a couple where one partner or both are sufferers or carriers of a single-gene disorder wish to have a child, they can do so through *in vitro* fertilization, which means they can then have a preimplantation genetic diagnosis to check whether the embryo has the genetic disorder.

Autosomal Dominant

Only one mutated copy of the gene will be necessary for a person to be affected by an autosomal dominant disorder. Each affected person usually has one affected parent. The chance a child will inherit the mutated gene is 50%. Autosomal dominant conditions sometimes have reduced penetrance, which means although only one mutated copy is needed, not all individuals who inherit that mutation go on to develop the disease. Examples of this type of disorder are Huntington's disease, neurofibromatosis type 1, neurofibromatosis type 2, Marfan syndrome, hereditary nonpolyposis colorectal cancer, hereditary multiple exostoses (a highly penetrant autosomal dominant disorder), Tuberous sclerosis, Von Willebrand disease, and acute intermittent porphyria. Birth defects are also called congenital anomalies.

Autosomal Recessive

Two copies of the gene must be mutated for a person to be affected by an autosomal recessive disorder. An affected person usually has unaffected parents who each carry a single copy of the mutated gene (and are referred to as carriers). Two unaffected people who each carry one copy of the mutated gene have a 25% risk with each pregnancy of having a child affected by the disorder. Examples of this type of disorder are Albinism, Medium-chain acyl-CoA dehydrogenase deficiency, cystic fibrosis, sickle-cell disease, Tay-Sachs disease, Niemann-Pick disease, spinal muscular atrophy, and Roberts syndrome. Certain other phenotypes, such as wet versus dry earwax, are also determined in an autosomal recessive fashion.

X-linked Dominant

X-linked dominant disorders are caused by mutations in genes on the X chromosome. Only a few disorders have this inheritance pattern, with a prime example being X-linked hypophos-phatemic rickets. Males and females are both affected in these disorders, with males typically being more severely affected than females. Some X-linked dominant conditions, such as Rett syndrome, incontinentia pigmenti type 2, and Aicardi syndrome, are usually fatal in males either *in utero* or shortly after birth, and are therefore predominantly seen in females. Exceptions to this finding are extremely rare cases in which boys with Klinefelter syndrome (47,XXY) also inherit an X-linked dominant condition and exhibit symptoms more similar to those of a female in terms of disease severity. The chance of passing on an X-linked dominant disorder differs between men and women. The sons of a man with an X-linked dominant disorder will all be unaffected (since they receive their father's Y chromosome), and his daughters will all inherit the condition. A woman with an X-linked dominant disorder has a 50% chance of having an affected fetus with each pregnancy, although it should be noted that in cases such as incontinentia pigmenti, only female offspring are generally viable. In addition, although these conditions do not alter fertility *per se*, individuals with Rett syndrome or Aicardi syndrome rarely reproduce.

X-linked Recessive

X-linked recessive conditions are also caused by mutations in genes on the X chromosome. Males are more frequently affected than females, and the chance of passing on the disorder differs between men and women. The sons of a man with an X-linked recessive disorder will not be affected, and his daughters will carry one copy of the mutated gene. A woman who is a carrier of an X-linked recessive disorder (X^RX^r) has a 50% chance of having sons who are affected and a 50% chance of having daughters who carry one copy of the mutated gene and are therefore carriers. X-linked recessive conditions include the serious diseases hemophilia A, Duchenne muscular dystrophy, and Lesch-Nyhan syndrome, as well as common and less serious conditions such as male pattern baldness and red-green color blindness. X-linked recessive conditions can sometimes manifest in females due to skewed X-inactivation or monosomy X (Turner syndrome).

Y-linked

Y-linked disorders, also called holandric disorders, are caused by mutations on the Y chromosome. These conditions display may only be transmitted from the heterogametic sex (e.g. male humans) to offspring of the same sex. More simply, this means that Y-linked disorders in humans can only be passed from men to their sons; females can never be affected because they do not possess Y-allosomes.

Y-linked disorders are exceedingly rare but the most well-known examples typically cause infertility. Reproduction in such conditions is only possible through the circumvention of infertility by medical intervention.

Mitochondrial

This type of inheritance, also known as maternal inheritance, applies to genes in mitochondrial DNA. Because only egg cells contribute mitochondria to the developing embryo, only mothers can pass on mitochondrial conditions to their children. An example of this type of disorder is Leber's hereditary optic neuropathy.

Many Genes

Genetic disorders may also be complex, multifactorial, or polygenic, meaning they are likely associated with the effects of multiple genes in combination with lifestyles and environmental factors. Multifactorial disorders include heart disease and diabetes. Although complex disorders often cluster in families, they do not have a clear-cut pattern of inheritance. This makes it difficult to determine a person's risk of inheriting or passing on these disorders. Complex disorders are also difficult to study and treat, because the specific factors that cause most of these disorders have not yet been identified. Studies which aim to identify the cause of complex disorders can use several methodological approaches to determine genotype-phenotype associations. One method, the genotype-first approach, starts by identifying genetic variants within patients and then determining the associated clinical manifestations. This is opposed to the more traditional phenotype-first approach, and may identify causal factors that have previously been obscured by clinical heterogeneity, penetrance, and expressivity.

On a pedigree, polygenic diseases do tend to "run in families", but the inheritance does not fit simple patterns as with Mendelian diseases. But this does not mean that the genes cannot eventually be located and studied. There is also a strong environmental component to many of them (e.g., blood pressure).

- asthma
- autoimmune diseases such as multiple sclerosis
- cancers
- ciliopathies
- cleft palate
- diabetes
- heart disease
- hypertension
- inflammatory bowel disease
- intellectual disability
- mood disorder
- obesity
- refractive error
- infertility

Diagnosis

Due to the wide range of genetic disorders that are presently known, diagnosis of a genetic disorder is widely varied and dependent of the disorder. Most genetic disorders are diagnosed at birth or during early childhood, however some, such as Huntington's disease, can escape detection until the patient is well into adulthood.

The basic aspects of a genetic disorder rests on the inheritance of genetic material. With an in depth family history, it is possible to anticipate possible disorders in children which direct medical professionals to specific tests depending on the disorder and allow parents the chance to prepare for potential lifestyle changes, anticipate the possibility of stillbirth, or contemplate termination. Prenatal diagnosis can detect the presence of characteristic abnormalities in fetal development through ultrasound, or detect the presence of characteristic substances via invasive procedures which involve inserting probes or needles into the uterus such as in amniocentesis.

Prognosis

Not all genetic disorders directly result in death, however there are no known cures for genetic disorders. Many genetic disorders affect stages of development such as Down's Syndrome. While others result in purely physical symptoms such as Muscular Dystrophy. Other disorders, such as Huntington's Disease show no signs until adulthood. During the active time of a genetic disorder, patients mostly rely on maintaining or slowing the degradation of quality of life and maintain patient autonomy. This includes physical therapy, pain management, and may include a selection of alternative medicine programs.

Treatment

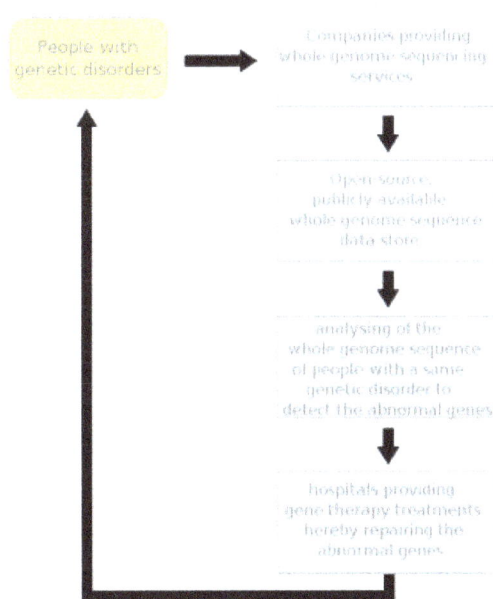

From personal genomics to gene therapy

The treatment of genetic disorders is an ongoing battle with over 1800 gene therapy clinical trials having been completed, are ongoing, or have been approved worldwide. Despite this, most treatment options revolve around treating the symptoms of the disorders in an attempt to improve patient quality of life.

Gene therapy refers to a form of treatment where a healthy gene is introduced to a patient. This should alleviate the defect caused by a faulty gene or slow the progression of disease. A major obstacle has been the delivery of genes to the appropriate cell, tissue, and organ affected by the disorder. How does one introduce a gene into the potentially trillions of cells which carry the defective copy? This question has been the roadblock between understanding the genetic disorder and correcting the genetic disorder.

Gene Therapy

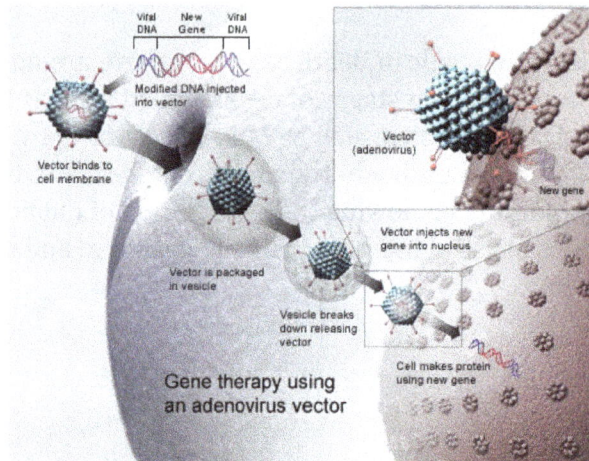

Gene therapy using an adenovirus vector. In some cases, the adenovirus will insert the new gene into a cell. If the treatment is successful, the new gene will make a functional protein to treat a disease.

Gene therapy is the therapeutic delivery of nucleic acid polymers into a patient's cells as a drug to treat disease.

The first attempt at modifying human DNA was performed in 1980 by Martin Cline, but the first successful and approved nuclear gene transfer in humans was performed in May 1989. The first therapeutic use of gene transfer as well as the first direct insertion of human DNA into the nuclear genome was performed by French Anderson in a trial starting in September 1990.

Between 1989 and February 2016, over 2,300 clinical trials had been conducted, more than half of them in phase I.

It should be noted that not all medical procedures that introduce alterations to a patient's genetic makeup can be considered gene therapy. Bone marrow transplantation and organ transplants in general have been found to introduce foreign DNA into patients. Gene therapy is defined by the precision of the procedure and the intention of direct therapeutic effects.

Background

Gene therapy was conceptualized in 1972, by authors who urged caution before commencing human gene therapy studies.

The first attempt, an unsuccessful one, at gene therapy (as well as the first case of medical transfer of foreign genes into humans not counting organ transplantation) was performed by Martin Cline on 10 July 1980. Cline claimed that one of the genes in his patients was active six months later, though he never published this data or had it verified and even if he is correct, it's unlikely it produced any significant beneficial effects treating beta-thalassemia.

After extensive research on animals throughout the 1980s and a 1989 bacterial gene tagging trial on humans, the first gene therapy widely accepted as a success was demonstrated in a trial that started on September 14, 1990, when Ashi DeSilva was treated for ADA-SCID.

The first somatic treatment that produced a permanent genetic change was performed in 1993.

This procedure was referred to sensationally and somewhat inaccurately in the media as a "three parent baby", though mtDNA is not the primary human genome and has little effect on an organism's individual characteristics beyond powering their cells.

Gene therapy is a way to fix a genetic problem at its source. The polymers are either translated into proteins, interfere with target gene expression, or possibly correct genetic mutations.

The most common form uses DNA that encodes a functional, therapeutic gene to replace a mutated gene. The polymer molecule is packaged within a "vector", which carries the molecule inside cells.

Early clinical failures led to dismissals of gene therapy. Clinical successes since 2006 regained researchers' attention, although as of 2014, it was still largely an experimental technique. These include treatment of retinal diseases Leber's congenital amaurosis and choroideremia, X-linked SCID, ADA-SCID, adrenoleukodystrophy, chronic lymphocytic leukemia (CLL), acute lymphocytic leukemia (ALL), multiple myeloma, haemophilia and Parkinson's disease. Between 2013 and April 2014, US companies invested over $600 million in the field.

The first commercial gene therapy, Gendicine, was approved in China in 2003 for the treatment of certain cancers. In 2011 Neovasculgen was registered in Russia as the first-in-class gene-therapy drug for treatment of peripheral artery disease, including critical limb ischemia. In 2012 Glybera, a treatment for a rare inherited disorder, became the first treatment to be approved for clinical use in either Europe or the United States after its endorsement by the European Commission.

Approaches

Following early advances in genetic engineering of bacteria, cells, and small animals, scientists started considering how to apply it to medicine. Two main approaches were considered – replacing or disrupting defective genes. Scientists focused on diseases caused by single-gene defects, such as cystic fibrosis, haemophilia, muscular dystrophy, thalassemia and sickle cell anemia. Glybera treats one such disease, caused by a defect in lipoprotein lipase.

DNA must be administered, reach the damaged cells, enter the cell and express/disrupt a protein. Multiple delivery techniques have been explored. The initial approach incorporated DNA into an engineered virus to deliver the DNA into a chromosome. Naked DNA approaches have also been explored, especially in the context of vaccine development.

Generally, efforts focused on administering a gene that causes a needed protein to be expressed. More recently, increased understanding of nuclease function has led to more direct DNA editing, using techniques such as zinc finger nucleases and CRISPR. The vector incorporates genes into

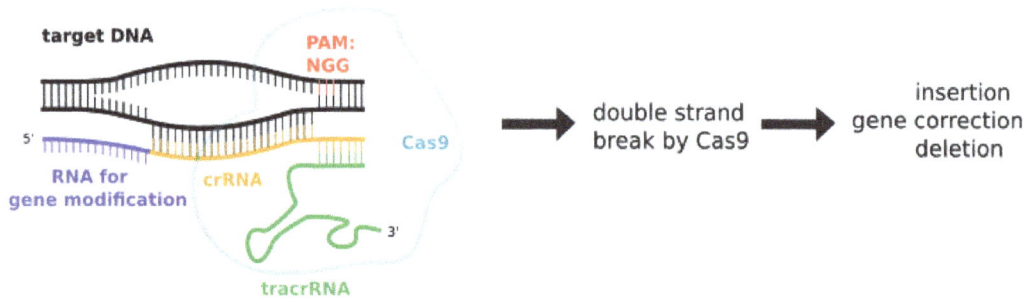

chromosomes. The expressed nucleases then knock out and replace genes in the chromosome. As of 2014 these approaches involve removing cells from patients, editing a chromosome and returning the transformed cells to patients.

Future of CRISPR-Cas 9

Gene editing has been a potential therapy for many genetic diseases. Targeted genome editing using nucleases provides a general method for inducing deletions or insertion. An earlier method for targeting relies on protein-DNA interactions; however, the most recent one, using CRISPR – associated protein 9 (Cas9), provides better specificity, simplicity, speed and pricing. The CRISPR System was first identified in single cell archaea (Prokaryotes). It is now widely used in different cell types and organisms including human cells (HEK293T, HeLa, iPSC), mouse, fruit fly, rice, wheat etc. CRISPR –Cas9 genome editing has potential applications in human gene therapy, screening drug targets, synthetic biology, agriculture, programmable RNA targeting, and viral gene disruption.

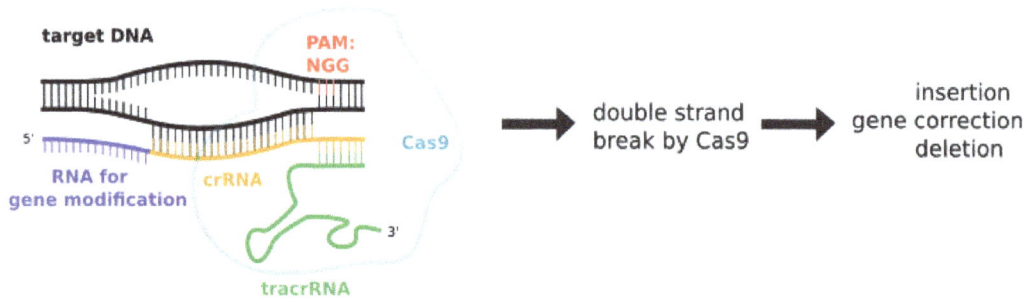

A duplex of crRNA and tracrRNA acts as guide RNA to introduce a specifically located gene modification based on the RNA 5' upstream of the crRNA. Cas9 binds the tracrRNA and needs a DNA binding sequence (5'NGG3'), which is called protospacer adjacent motif (PAM). After binding, Cas9 introduces a DNA double strand break, which is then followed by gene modification via homologous recombination (HDR) or non-homologous end joining (NHEJ).

Other technologies employ antisense, small interfering RNA and other DNA. To the extent that these technologies do not alter DNA, but instead directly interact with molecules such as RNA, they are not considered "gene therapy" per se.

Cell Types

Somatic Cell

In somatic cell gene therapy (SCGT), the therapeutic genes are transferred into any cell other than a gamete, germ cell, gametocyte or undifferentiated stem cell. Any such modifications affect the individual patient only, and are not inherited by offspring. Somatic gene therapy represents mainstream basic and clinical research, in which therapeutic DNA (either integrated in the genome or as an external episome or plasmid) is used to treat disease.

Over 600 clinical trials utilizing SCGT are underway in the US. Most focus on severe genetic disorders, including immunodeficiencies, haemophilia, thalassaemia and cystic fibrosis. Such single gene disorders are good candidates for somatic cell therapy. The complete correction of a genetic disorder or the replacement of multiple genes is not yet possible. Only a few of the trials are in the advanced stages.

Germline

In germline gene therapy (GGT), germ cells (sperm or eggs) are modified by the introduction of functional genes into their genomes. Modifying a germ cell causes all the organism's cells to contain the modified gene. The change is therefore heritable and passed on to later generations. Australia, Canada, Germany, Israel, Switzerland and the Netherlands prohibit GGT for application in human beings, for technical and ethical reasons, including insufficient knowledge about possible risks to future generations and higher risks versus SCGT. The US has no federal controls specifically addressing human genetic modification (beyond FDA regulations for therapies in general).

Vectors

The delivery of DNA into cells can be accomplished by multiple methods. The two major classes are recombinant viruses (sometimes called biological nanoparticles or viral vectors) and naked DNA or DNA complexes (non-viral methods).

Viruses

In order to replicate, viruses introduce their genetic material into the host cell, tricking the host's cellular machinery into using it as blueprints for viral proteins. Scientists exploit this by substituting a virus's genetic material with therapeutic DNA. (The term 'DNA' may be an oversimplification, as some viruses contain RNA, and gene therapy could take this form as well.) A number of viruses have been used for human gene therapy, including retrovirus, adenovirus, lentivirus, herpes simplex, vaccinia and adeno-associated virus. Like the genetic material (DNA or RNA) in viruses, therapeutic DNA can be designed to simply serve as a temporary blueprint that is degraded naturally or (at least theoretically) to enter the host's genome, becoming a permanent part of the host's DNA in infected cells.

Non-viral

Non-viral methods present certain advantages over viral methods, such as large scale production and low host immunogenicity. However, non-viral methods initially produced lower levels of transfection and gene expression, and thus lower therapeutic efficacy. Later technology remedied this deficiency.

Methods for non-viral gene therapy include the injection of naked DNA, electroporation, the gene gun, sonoporation, magnetofection, the use of oligonucleotides, lipoplexes, dendrimers, and inorganic nanoparticles.

Hurdles

Some of the unsolved problems include:

- Short-lived nature – Before gene therapy can become a permanent cure for a condition, the therapeutic DNA introduced into target cells must remain functional and the cells containing the therapeutic DNA must be stable. Problems with integrating therapeutic DNA into the genome and the rapidly dividing nature of many cells prevent it from achieving long-term benefits. Patients require multiple treatments.

- Immune response – Any time a foreign object is introduced into human tissues, the immune system is stimulated to attack the invader. Stimulating the immune system in a way that reduces gene therapy effectiveness is possible. The immune system's enhanced response to viruses that it has seen before reduces the effectiveness to repeated treatments.

- Problems with viral vectors – Viral vectors carry the risks of toxicity, inflammatory responses, and gene control and targeting issues.

- Multigene disorders – Some commonly occurring disorders, such as heart disease, high blood pressure, Alzheimer's disease, arthritis, and diabetes, are affected by variations in multiple genes, which complicate gene therapy.

- Some therapies may breach the Weismann barrier (between soma and germ-line) protecting the testes, potentially modifying the germline, falling afoul of regulations in countries that prohibit the latter practice.

- Insertional mutagenesis – If the DNA is integrated in a sensitive spot in the genome, for example in a tumor suppressor gene, the therapy could induce a tumor. This has occurred in clinical trials for X-linked severe combined immunodeficiency (X-SCID) patients, in which hematopoietic stem cells were transduced with a corrective transgene using a retrovirus, and this led to the development of T cell leukemia in 3 of 20 patients. One possible solution is to add a functional tumor suppressor gene to the DNA to be integrated. This may be problematic since the longer the DNA is, the harder it is to integrate into cell genomes. CRISPR technology allows researchers to make much more precise genome changes at exact locations.

- Cost – Alipogene tiparvovec or Glybera, for example, at a cost of $1.6 million per patient, was reported in 2013 to be the world's most expensive drug.

Deaths

Three patients' deaths have been reported in gene therapy trials, putting the field under close scrutiny. The first was that of Jesse Gelsinger in 1999. One X-SCID patient died of leukemia in 2003. In 2007, a rheumatoid arthritis patient died from an infection; the subsequent investigation concluded that the death was not related to gene therapy.

History

1970s and Earlier

In 1972 Friedmann and Roblin authored a paper in *Science* titled "Gene therapy for human genetic disease?" Rogers (1970) was cited for proposing that *exogenous good DNA* be used to replace the defective DNA in those who suffer from genetic defects.

1980s

In 1984 a retrovirus vector system was designed that could efficiently insert foreign genes into mammalian chromosomes.

1990s

The first approved gene therapy clinical research in the US took place on 14 September 1990, at the National Institutes of Health (NIH), under the direction of William French Anderson. Four-year-old Ashanti DeSilva received treatment for a genetic defect that left her with ADA-SCID, a severe immune system deficiency. The effects were temporary, but successful.

Cancer gene therapy was introduced in 1992/93 (Trojan et al. 1993). The treatment of glioblastoma multiforme, the malignant brain tumor whose outcome is always fatal, was done using a vector expressing antisense IGF-I RNA (clinical trial approved by NIH n° 1602, and FDA in 1994). This therapy also represents the beginning of cancer immunogene therapy, a treatment which proves to be effective due to the anti-tumor mechanism of IGF-I antisense, which is related to strong immune and apoptotic phenomena.

In 1992 Claudio Bordignon, working at the Vita-Salute San Raffaele University, performed the first gene therapy procedure using hematopoietic stem cells as vectors to deliver genes intended to correct hereditary diseases. In 2002 this work led to the publication of the first successful gene therapy treatment for adenosine deaminase-deficiency (SCID). The success of a multi-center trial for treating children with SCID (severe combined immune deficiency or "bubble boy" disease) from 2000 and 2002, was questioned when two of the ten children treated at the trial's Paris center developed a leukemia-like condition. Clinical trials were halted temporarily in 2002, but resumed after regulatory review of the protocol in the US, the United Kingdom, France, Italy and Germany.

In 1993 Andrew Gobea was born with SCID following prenatal genetic screening. Blood was removed from his mother's placenta and umbilical cord immediately after birth, to acquire stem cells. The allele that codes for adenosine deaminase (ADA) was obtained and inserted into a retrovirus. Retroviruses and stem cells were mixed, after which the viruses inserted the gene into the stem cell chromosomes. Stem cells containing the working ADA gene were injected into Andrew's blood. Injections of the ADA enzyme were also given weekly. For four years T cells (white blood cells), produced by stem cells, made ADA enzymes using the ADA gene. After four years more treatment was needed.

Jesse Gelsinger's death in 1999 impeded gene therapy research in the US. As a result, the FDA suspended several clinical trials pending the reevaluation of ethical and procedural practices.

2000s

The modified cancer gene therapy strategy of antisense IGF-I RNA (NIH n° 1602) using antisense / triple helix anti IGF-I approach was registered in 2002 by Wiley gene therapy clinical trial - n° 635 and 636. The approach has shown promising results in the treatment of six different malignant tumors: glioblastoma, cancers of liver, colon, prostate, uterus and ovary (Collaborative NATO Science Programme on Gene Therapy USA, France, Poland n° LST 980517 conducted by J. Trojan) (Trojan et al., 2012). This anti–gene antisense/triple helix therapy has proven to be efficient, due to the mechanism stopping simultaneously IGF-I expression on translation and transcription levels, strengthening anti-tumor immune and apoptotic phenomena.

2002

Sickle-cell disease can be treated in mice. The mice – which have essentially the same defect that causes human cases – used a viral vector to induce production of fetal hemoglobin (HbF), which normally ceases to be produced shortly after birth. In humans, the use of hydroxyurea to stimulate the production of HbF temporarily alleviates sickle cell symptoms. The researchers demonstrated this treatment to be a more permanent means to increase therapeutic HbF production.

A new gene therapy approach repaired errors in messenger RNA derived from defective genes. This technique has the potential to treat thalassaemia, cystic fibrosis and some cancers.

Researchers created liposomes 25 nanometers across that can carry therapeutic DNA through pores in the nuclear membrane.

2003

In 2003 a research team inserted genes into the brain for the first time. They used liposomes coated in a polymer called polyethylene glycol, which, unlike viral vectors, are small enough to cross the blood–brain barrier.

Short pieces of double-stranded RNA (short, interfering RNAs or siRNAs) are used by cells to degrade RNA of a particular sequence. If a siRNA is designed to match the RNA copied from a faulty gene, then the abnormal protein product of that gene will not be produced.

Gendicine is a cancer gene therapy that delivers the tumor suppressor gene p53 using an engineered adenovirus. In 2003, it was approved in China for the treatment of head and neck squamous cell carcinoma.

2006

In March researchers announced the successful use of gene therapy to treat two adult patients for X-linked chronic granulomatous disease, a disease which affects myeloid cells and damages the immune system. The study is the first to show that gene therapy can treat the myeloid system.

In May a team reported a way to prevent the immune system from rejecting a newly delivered gene. Similar to organ transplantation, gene therapy has been plagued by this problem. The immune system normally recognizes the new gene as foreign and rejects the cells carrying it. The research utilized a newly uncovered network of genes regulated by molecules known as microRNAs. This natural function selectively obscured their therapeutic gene in immune system cells and protected it from discovery. Mice infected with the gene containing an immune-cell microRNA target sequence did not reject the gene.

In August scientists successfully treated metastatic melanoma in two patients using killer T cells genetically retargeted to attack the cancer cells.

In November researchers reported on the use of VRX496, a gene-based immunotherapy for the treatment of HIV that uses a lentiviral vector to deliver an antisense gene against the HIV envelope. In a phase I clinical trial, five subjects with chronic HIV infection who had failed to respond to at least two antiretroviral regimens were treated. A single intravenous infusion of autologous CD4 T cells genetically modified with VRX496 was well tolerated. All patients had stable or decreased viral load; four of the five patients had stable or increased CD4 T cell counts. All five patients had stable or increased immune response to HIV antigens and other pathogens. This was the first evaluation of a lentiviral vector administered in a US human clinical trial.

2007

In May researchers announced the first gene therapy trial for inherited retinal disease. The first operation was carried out on a 23-year-old British male, Robert Johnson, in early 2007.

2008

Leber's congenital amaurosis is an inherited blinding disease caused by mutations in the RPE65 gene. The results of a small clinical trial in children were published in April. Delivery of recombinant adeno-associated virus (AAV) carrying RPE65 yielded positive results. In May two more groups reported positive results in independent clinical trials using gene therapy to treat the condition. In all three clinical trials, patients recovered functional vision without apparent side-effects.

2009

In September researchers were able to give trichromatic vision to squirrel monkeys. In November 2009, researchers halted a fatal genetic disorder called adrenoleukodystrophy in two children using a lentivirus vector to deliver a functioning version of ABCD1, the gene that is mutated in the disorder.

2010s

2010

An April paper reported that gene therapy addressed achromatopsia (color blindness) in dogs by targeting cone photoreceptors. Cone function and day vision were restored for at least 33 months in two young specimens. The therapy was less efficient for older dogs.

In September it was announced that an 18-year-old male patient in France with beta-thalassemia major had been successfully treated. Beta-thalassemia major is an inherited blood disease in which beta haemoglobin is missing and patients are dependent on regular lifelong blood transfusions. The technique used a lentiviral vector to transduce the human ß-globin gene into purified blood and marrow cells obtained from the patient in June 2007. The patient's haemoglobin levels were stable at 9 to 10 g/dL. About a third of the hemoglobin contained the form introduced by the viral vector and blood transfusions were not needed. Further clinical trials were planned. Bone marrow transplants are the only cure for thalassemia, but 75% of patients do not find a matching donor.

2011

In 2007 and 2008, a man was cured of HIV by repeated Hematopoietic stem cell transplantation with double-delta-32 mutation which disables the CCR5 receptor. This cure was accepted by the medical community in 2011. It required complete ablation of existing bone marrow, which is very debilitating.

In August two of three subjects of a pilot study were confirmed to have been cured from chronic lymphocytic leukemia (CLL). The therapy used genetically modified T cells to attack cells that expressed the CD19 protein to fight the disease. In 2013, the researchers announced that 26 of 59 patients had achieved complete remission and the original patient had remained tumor-free.

Human HGF plasmid DNA therapy of cardiomyocytes is being examined as a potential treatment for coronary artery disease as well as treatment for the damage that occurs to the heart after myocardial infarction.

In 2011 Neovasculgen was registered in Russia as the first-in-class gene-therapy drug for treatment of peripheral artery disease, including critical limb ischemia; it delivers the gene encoding for VEGF. Neovasculogen is a plasmid encoding the CMV promoter and the 165 amino acid form of VEGF.

2012

The FDA approved Phase 1 clinical trials on thalassemia major patients in the US for 10 participants in July. The study was expected to continue until 2015.

In July 2012, the European Medicines Agency recommended approval of a gene therapy treatment for the first time in either Europe or the United States. The treatment used Alipogene tiparvovec (Glybera) to compensate for lipoprotein lipase deficiency, which can cause severe pancreatitis. The recommendation was endorsed by the European Commission in November 2012 and commercial rollout began in late 2014.

In December 2012, it was reported that 10 of 13 patients with multiple myeloma were in remission "or very close to it" three months after being injected with a treatment involving genetically engineered T cells to target proteins NY-ESO-1 and LAGE-1, which exist only on cancerous myeloma cells.

2013

In March researchers reported that three of five subjects who had acute lymphocytic leukemia (ALL) had been in remission for five months to two years after being treated with genetically modified T cells which attacked cells with CD19 genes on their surface, i.e. all B-cells, cancerous or not. The researchers believed that the patients' immune systems would make normal T-cells and B-cells after a couple of months. They were also given bone marrow. One patient relapsed and died and one died of a blood clot unrelated to the disease.

Following encouraging Phase 1 trials, in April, researchers announced they were starting Phase 2 clinical trials (called CUPID2 and SERCA-LVAD) on 250 patients at several hospitals to combat heart disease. The therapy was designed to increase the levels of SERCA2, a protein in heart muscles, improving muscle function. The FDA granted this a Breakthrough Therapy Designation to accelerate the trial and approval process. In 2016 it was reported that no improvement was found from the CUPID 2 trial.

In July researchers reported promising results for six children with two severe hereditary diseases had been treated with a partially deactivated lentivirus to replace a faulty gene and after 7–32 months. Three of the children had metachromatic leukodystrophy, which causes children to lose cognitive and motor skills. The other children had Wiskott-Aldrich syndrome, which leaves them to open to infection, autoimmune diseases and cancer. Follow up trials with gene therapy on another six children with Wiskott-Aldrich syndrome were also reported as promising.

In October researchers reported that two children born with adenosine deaminase severe combined immunodeficiency disease (ADA-SCID) had been treated with genetically engineered stem cells 18 months previously and that their immune systems were showing signs of full recovery. Another three children were making progress. In 2014 a further 18 children with ADA-SCID were cured by gene therapy. ADA-SCID children have no functioning immune system and are sometimes known as "bubble children."

Also in October researchers reported that they had treated six haemophilia sufferers in early 2011 using an adeno-associated virus. Over two years later all six were producing clotting factor.

Data from three trials on Topical cystic fibrosis transmembrane conductance regulator gene therapy were reported to not support its clinical use as a mist inhaled into the lungs to treat cystic fibrosis patients with lung infections.

2014

In January researchers reported that six choroideremia patients had been treated with adeno-associated virus with a copy of REP1. Over a six-month to two-year period all had improved their sight. By 2016, 32 patients had been treated with positive results and researchers were hopeful the treatment would be long-lasting. Choroideremia is an inherited genetic eye disease with no approved treatment, leading to loss of sight.

In March researchers reported that 12 HIV patients had been treated since 2009 in a trial with a genetically engineered virus with a rare mutation (CCR5 deficiency) known to protect against HIV with promising results.

Clinical trials of gene therapy for sickle cell disease were started in 2014 although one review failed to find any such trials.

2015

In February LentiGlobin BB305, a gene therapy treatment undergoing clinical trials for treatment of beta thalassemia gained FDA "breakthrough" status after several patients were able to forgo the frequent blood transfusions usually required to treat the disease.

In March researchers delivered a recombinant gene encoding a broadly neutralizing antibody into monkeys infected with simian HIV; the monkeys' cells produced the antibody, which cleared them of HIV. The technique is named immunoprophylaxis by gene transfer (IGT). Animal tests for antibodies to ebola, malaria, influenza and hepatitis are underway.

In March scientists, including an inventor of CRISPR, urged a worldwide moratorium on germline gene therapy, writing "scientists should avoid even attempting, in lax jurisdictions, germline genome modification for clinical application in humans" until the full implications "are discussed among scientific and governmental organizations".

Also in 2015 Glybera was approved for the German market.

In October, researchers announced that they had treated a baby girl, Layla Richards, with an experimental treatment using donor T-cells genetically engineered to attack cancer cells. Two months after the treatment she was still free of her cancer (a highly aggressive form of acute lymphoblastic leukaemia [ALL]). Children with highly aggressive ALL normally have a very poor prognosis and Layla's disease had been regarded as terminal before the treatment.

In December, scientists of major world academies called for a moratorium on inheritable human genome edits, including those related to CRISPR-Cas9 technologies but that basic research including embryo gene editing should continue.

2016

In April the Committee for Medicinal Products for Human Use of the European Medicines Agency endorsed a gene therapy treatment called Strimvelis and recommended it be approved. This treats children born with ADA-SCID and who have no functioning immune system - sometimes called the "bubble baby" disease. This would be the second gene therapy treatment to be approved in Europe.

Fertility

Gene Therapy techniques have the potential to provide alternative treatments for those with infertility. Recently, successful experimentation on mice has proven that fertility can be restored by using the gene therapy method, CRISPR. Spermatogenical stem cells from another organism were transplanted into the testes of an infertile male mouse. The stem cells re-established spermatogenesis and fertility.

Gene Doping

Athletes might adopt gene therapy technologies to improve their performance. Gene doping is not known to occur, but multiple gene therapies may have such effects. Kayser et al. argue that gene doping could level the playing field if all athletes receive equal access. Critics claim that any therapeutic intervention for non-therapeutic/enhancement purposes compromises the ethical foundations of medicine and sports.

Human Genetic Engineering

Genetic engineering could be used to change physical appearance, metabolism, and even improve physical capabilities and mental faculties such as memory and intelligence. Ethical claims about germline engineering include beliefs that every fetus has a right to remain genetically unmodified, that parents hold the right to genetically modify their offspring, and that every child has the right to be born free of preventable diseases. For adults, genetic engineering could be seen as another enhancement technique to add to diet, exercise, education, cosmetics and plastic surgery. Another theorist claims that moral concerns limit but do not prohibit germline engineering.

Possible regulatory schemes include a complete ban, provision to everyone, or professional self-regulation. The American Medical Association's Council on Ethical and Judicial Affairs stated that "genetic interventions to enhance traits should be considered permissible only in severely restricted situations: (1) clear and meaningful benefits to the fetus or child; (2) no trade-off with other characteristics or traits; and (3) equal access to the genetic technology, irrespective of income or other socioeconomic characteristics."

As early in the history of biotechnology as 1990, there have been scientists opposed to attempts to modify the human germline using these new tools, and such concerns have continued as technology progressed. With the advent of new techniques like CRISPR, in March 2015 a group of scientists urged a worldwide moratorium on clinical use of gene editing technologies to edit the human genome in a way that can be inherited. In April 2015, researchers sparked controversy when they reported results of basic research to edit the DNA of non-viable human embryos using CRISPR.

Regulations

Regulations covering genetic modification are part of general guidelines about human-involved biomedical research.

The Helsinki Declaration (Ethical Principles for Medical Research Involving Human Subjects) was amended by the World Medical Association's General Assembly in 2008. This document provides principles physicians and researchers must consider when involving humans as research subjects. The Statement on Gene Therapy Research initiated by the Human Genome Organization (HUGO) in 2001 provides a legal baseline for all countries. HUGO's document emphasizes human freedom and adherence to human rights, and offers recommendations for somatic gene therapy, including the importance of recognizing public concerns about such research.

United States

No federal legislation lays out protocols or restrictions about human genetic engineering. This subject is governed by overlapping regulations from local and federal agencies, including the Department of Health and Human Services, the FDA and NIH's Recombinant DNA Advisory Committee. Researchers seeking federal funds for an investigational new drug application, (commonly the case for somatic human genetic engineering), must obey international and federal guidelines for the protection of human subjects.

NIH serves as the main gene therapy regulator for federally funded research. Privately funded research is advised to follow these regulations. NIH provides funding for research that develops or enhances genetic engineering techniques and to evaluate the ethics and quality in current research. The NIH maintains a mandatory registry of human genetic engineering research protocols that includes all federally funded projects.

An NIH advisory committee published a set of guidelines on gene manipulation. The guidelines discuss lab safety as well as human test subjects and various experimental types that involve genetic changes. Several sections specifically pertain to human genetic engineering, including Section III-C-1. This section describes required review processes and other aspects when seeking approval to begin clinical research involving genetic transfer into a human patient. The protocol for a gene therapy clinical trial must be approved by the NIH's Recombinant DNA Advisory Committee prior to any clinical trial beginning; this is different from any other kind of clinical trial.

As with other kinds of drugs, the FDA regulates the quality and safety of gene therapy products and supervises how these products are used clinically. Therapeutic alteration of the human genome falls under the same regulatory requirements as any other medical treatment. Research involving human subjects, such as clinical trials, must be reviewed and approved by the FDA and an Institutional Review Board.

References

- Barton NH, Briggs DE, Eisen JA, Goldstein DB, Patel NH (2007). Evolution. Cold Spring Harbor, NY: Cold Spring Harbor Laboratory Press. ISBN 0-87969-684-2.

- Griffiths, Anthony J.F.; Wessler, Susan R.; Carroll, Sean B.; Doebley, John (2012). "2: Single-Gene Inheritance". Introduction to Genetic Analysis (10th ed.). New York: W.H. Freeman and Company. p. 57. ISBN 978-1-4292-2943-2.

- Griffiths, Anthony J.F.; Wessler, Susan R.; Carroll, Sean B.; Doebley, John (2012). Introduction to Genetic Analysis (10th ed.). New York: W.H. Freeman and Company. p. 58. ISBN 978-1-4292-2943-2.

- Milunsky, edited by Aubrey (2004). Genetic disorders and the fetus : diagnosis, prevention, and treatment (5th ed.). Baltimore: Johns Hopkins University Press. ISBN 0801879280.

- Pollack, Andrew (2 June 2016). "Scientists Announce HGP-Write, Project to Synthesize the Human Genome". New York Times. Retrieved 2 June 2016.

- Ghosh, Pallab (28 April 2016). "Gene therapy reverses sight loss and is long-lasting". BBC News, Science & Environment. Retrieved 29 April 2016.

- Fernàndez-Ruiz, Irene (4 February 2016). "No improvement in outcomes with gene therapy for heart failure". Nature Reviews Cardiology. 13: 122–123. doi:10.1038/nrcardio.2016.14. Retrieved 29 April 2016.

- "Summary of opinion1 (initial authorisation) Strimvelis" (PDF). European Medicines Agency. 1 April 2016. pp. 1–2. Retrieved 13 April 2016.

- Priya Moorjani, Kumarasamy Thangaraj, Nick Patterson, Mark Lipson, Po-Ru Loh, Periyasamy Govindaraj, Bonnie Berger, David Reich, Lalji Singh (5 September 2013). "Genetic Evidence for Recent Population Mixture in India" (PDF). American Journal of Human Genetics. 93 (3): 422–438. doi:10.1016/j.ajhg.2013.07.006. PMC 3769933. PMID 23932107. Retrieved 17 May 2015.

- Rakesh Tamang, Lalji Singh, Kumarasamy Thangaraj (November 2012). "Complex genetic origin of Indian populations and its implications" (PDF). Journal of Biosciences. 37 (5): 911–919. doi:10.1007/s12038-012-9256-9. Retrieved 17 May 2015.

- "Gene Therapy Turns Several Leukemia Patients Cancer Free. Will It Work for Other Cancers, Too?". Singularity Hub. Retrieved 7 January 2014.

- "Chiesi and uniQure delay Glybera launch to add data". Biotechnology. The Pharma Letter. 4 August 2014. Retrieved 28 August 2014.

- BURGER, LUDWIG; HIRSCHLER, BEN (November 26, 2014). "First gene therapy drug sets million-euro price record". Reuters. Retrieved March 2015.

5

Evolution of Genetics

This chapter chronicles the history of genetics from the days of Mendelian or classical genetics to the current genetic research in cancer therapy. The reader is provided with in-depth information about the history of genetics from its humble beginnings at a monastery. The timeline of genetics illustrates the immense and rapid advancement the field has experienced and discusses its future.

Human genetic clustering is the degree to which human genetic variation can be partitioned into a small number of groups or clusters. A leading method of analysis uses mathematical cluster analysis of the degree of similarity of genetic data between individuals and groups in order to infer population structures and assign individuals to hypothesized ancestral groups. These groupings in turn often, but not always, correspond with the individuals' self-identified geographical ancestry. A similar analysis can be done using principal components analysis, and several recent studies deploy both methods.

Analysis of genetic clustering examines the degree to which regional groups differ genetically, the categorization of individuals into clusters, and what can be learned about human ancestry from this data. There is broad scientific agreement that a relatively small fraction of human genetic variation occurs between populations, continents, or clusters. Researchers of genetic clustering differ, however, on whether genetic variation is principally clinal or whether clusters inferred mathematically are important and scientifically useful.

Analysis of Human Genetic Variation

Quantifying Variation

One of the underlying questions regarding the distribution of human genetic diversity is related to the degree to which genes are shared between the observed clusters. It has been observed repeatedly that the majority of variation observed in the global human population is found within populations. This variation is usually calculated using Sewall Wright's Fixation index (F_{ST}), which is an estimate of between to within group variation. The degree of human genetic variation is a little different depending upon the gene type studied, but in general it is common to claim that ~85% of genetic variation is found within groups, ~6–10% between groups within the same continent and ~6–10% is found between continental groups. These average numbers, however, do not mean that every population harbors an equal amount of diversity. In fact, some human populations contain far more genetic diversity than others, which is consistent with the likely African origin of modern humans. Therefore, populations outside of Africa may have undergone serial founder effects that limited their genetic diversity.

The F_{ST} statistic has been come under criticism by A. W. F. Edwards and Jeffrey Long and Rick

Kittles. British statistician and evolutionary biologist A. W. F. Edwards faulted Lewontin's methodology for basing his conclusions on simple comparison of genes and rather on a more complex structure of gene frequencies. Long and Kittles' objection is also methodological: according to them the F_{ST} is based on a faulty underlying assumptions that all populations contain equally genetic diverse members and that continental groups diverged at the same time. Sarich and Miele have also argued that estimates of genetic difference between individuals of different populations understate differences between groups because they fail to take into account human diploidy.

Keith Hunley, Graciela Cabana, and Jeffrey Long created a revised statistical model to account for unequally divergent population lineages and local populations with differing degrees of diversity. Their 2015 paper applies this model to the Human Genome Diversity Project sample of 1,037 individuals in 52 populations. They found that least diverse population examined, the Surui, "harbors nearly 60% of the total species' diversity." Long and Kittles had noted earlier that the Sokoto people of Africa contains virtually all of human genetic diversity. Their analysis also found that non-African populations are a taxonomic subgroup of African populations, that "some African populations are equally related to other African populations and to non-African populations," and that "outside of Africa, regional groupings of populations are nested inside one another, and many of them are not monophyletic."

Similarity of Group Members

Multiple studies since 1972 have backed up the claim that, "The average proportion of genetic differences between individuals from different human populations only slightly exceeds that between unrelated individuals from a single population."

Percentage similarity between two individuals from different clusters when 377 microsatellite markers are considered.			
x	**Africans**	**Europeans**	**Asians**
Europeans	36.5	—	—
Asians	35.5	38.3	—
Indigenous Americans	26.1	33.4	35

Edwards (2003) claims, "It is not true, as *Nature* claimed, that 'two random individuals from any one group are almost as different as any two random individuals from the entire world'" and Risch *et al.* (2002) state "Two Caucasians are more similar to each other genetically than a Caucasian and an Asian." However Bamshad et al. (2004) used the data from Rosenberg et al. (2002) to investigate the extent of genetic differences between individuals within continental groups relative to genetic differences between individuals between continental groups. They found that though these individuals could be classified very accurately to continental clusters, there was a significant degree of genetic overlap on the individual level, to the extent that, using 377 loci, individual Europeans were about 38% of the time more genetically similar to East Asians than to other Europeans.

Witherspoon et al. (2007) have argued that even when individuals can be reliably assigned to specific population groups, it may still be possible for two randomly chosen individuals from different populations/clusters to be more similar to each other than to a randomly chosen member of their own cluster. Witherspoon et al. conclude that "caution should be used when using geographic or genetic ancestry to make inferences about individual phenotypes." A study of three completely genotyped individuals, white American scientists James Watson and Craig Venter, and Korean scientist Seong-Jin Kim found that the two white scientists have fewer genetic variations (single nucleotide polymorphisms or SNPs) in common than either shares with Kim.

Genetic Cluster Studies

Genetic structure studies are carried out using statistical computer programs designed to find clusters of genetically similar individuals within a sample of individuals. Studies such as those by Risch and Rosenberg use a computer program called STRUCTURE to find human populations (gene clusters). It is a statistical program that works by placing individuals into one of an arbitrary number of clusters based on their overall genetic similarity, many possible pairs of clusters are tested per individual to generate multiple clusters. The basis for these computations are data describing a large number of single nucleotide polymorphisms (SNPs), genetic insertions and deletions (indels), microsatellite markers (or short tandem repeats, STRs) as they appear in each sampled individual. Cluster analysis divides a dataset into any prespecified number of clusters.

These clusters are based on multiple genetic markers that are often shared between different human populations even over large geographic ranges. The notion of a genetic cluster is that people within the cluster share on average similar allele frequencies to each other than to those in other clusters. (A. W. F. Edwards, 2003 but infobox "Multi Locus Allele Clusters") In a test of idealised populations, the computer programme STRUCTURE was found to consistently under-estimate the numbers of populations in the data set when high migration rates between populations and slow mutation rates (such as single-nucleotide polymorphisms) were considered. In 2004, Lynn Jorde and Steven Wooding argued that "Analysis of many loci now yields reasonably accurate estimates of genetic similarity among individuals, rather than populations. Clustering of individuals is correlated with geographic origin or ancestry."

A number of genetic cluster studies have been conducted since 2002, including the following:

Authors	Year	Title	Sample size / number of populations sampled	Sample	Markers
Rosenberg et al.	2002	Genetic Structure of Human Populations	1056 / 52	Human Genome Diversity Project (HGDP-CEPH)	377 STRs
Serre & Pääbo	2004	Worldwide Human Relationships Inferred from Genome-Wide Patterns of Variation	89 / 15	a: HGDP	20 STRs
			90 / geographically distributed individuals	b: Jorde 1997	
Rosenberg et al.	2005	Clines, Clusters, and the Effect of Study Design on the Inference of Human Population Structure	1056 / 52	Human Genome Diversity Project (HGDP-CEPH)	783 STRs + 210 indels
Li et al.	2008	Worldwide Human Relationships Inferred from Genome-Wide Patterns of Variation	938 / 51	Human Genome Diversity Project (HGDP-CEPH)	650,000 SNPs
Tishkoff et al.	2009	The Genetic Structure and History of Africans and African Americans	~3400 / 185	HGDP-CEPH *plus* 133 additional African populations and Indian individuals	1327 STRs + indels

Xing et al.	2010	Toward a more uniform sampling of human genetic diversity: A survey of worldwide populations by high-density genotyping	850 / 40	HapMap *plus* 296 individuals	250,000 SNPs

In a 2005 paper, Rosenberg and his team acknowledged that findings of a study on human population structure are highly influenced by the way the study is designed. They reported that the number of loci, the sample size, the geographic dispersion of the samples and assumptions about allele-frequency correlation all have an effect on the outcome of the study.

In a review of studies of human genome diversity, Guido Barbujani and colleagues note that various cluster studies have identified different numbers of clusters with different boundaries. They write that discordant patterns of genetic variation and high within-population genetic diversity "make[] it difficult, or impossible, to define, once and for good, the main genetic clusters of humankind."

Clusters by Rosenberg et al. (2002, 2005)

A major finding of Rosenberg and colleagues (2002) was that when five clusters were generated by the program (specified as K=5), "clusters corresponded largely to major geographic regions." Specifically, the five clusters corresponded to Africa, Europe plus the Middle East plus Central and South Asia, East Asia, Oceania, and the Americas. The study also confirmed prior analyses by showing that, "Within-population differences among individuals account for 93 to 95% of genetic variation; differences among major groups constitute only 3 to 5%."

Rosenberg and colleagues (2005) have argued, based on cluster analysis, that populations do not always vary continuously and a population's genetic structure is consistent if enough genetic markers (and subjects) are included. "Examination of the relationship between genetic and geographic distance supports a view in which the clusters arise not as an artifact of the sampling scheme, but from small discontinuous jumps in genetic distance for most population pairs on opposite sides of geographic barriers, in comparison with genetic distance for pairs on the same side. Thus, analysis of the 993-locus dataset corroborates our earlier results: if enough markers are used with a sufficiently large worldwide sample, individuals can be partitioned into genetic clusters that match major geographic subdivisions of the globe, with some individuals from intermediate geographic locations having mixed membership in the clusters that correspond to neighboring regions." They also wrote, regarding a model with five clusters corresponding to Africa, Eurasia (Europe, Middle East, and Central/South Asia), East Asia, Oceania, and the Americas: "For population pairs from the same cluster, as geographic distance increases, genetic distance increases in a linear manner, consistent with a clinal population structure. However, for pairs from different clusters, genetic distance is generally larger than that between intracluster pairs that have the same geographic distance. For example, genetic distances for population pairs with one population in Eurasia and the other in East Asia are greater than those for pairs at equivalent geographic distance within Eurasia or within East Asia. Loosely speaking, it is these small discontinuous jumps in genetic distance—across oceans, the Himalayas, and the Sahara—that provide the basis for the ability of STRUCTURE to identify clusters that correspond to geographic regions".

Rosenberg stated that their findings "should not be taken as evidence of our support of any particular concept of biological race (...). Genetic differences among human populations derive mainly from gradations in allele frequencies rather than from distinctive 'diagnostic' genotypes." The study's overall results confirmed that genetic difference within populations is between 93 and 95%. Only 5% of genetic variation is found between groups.

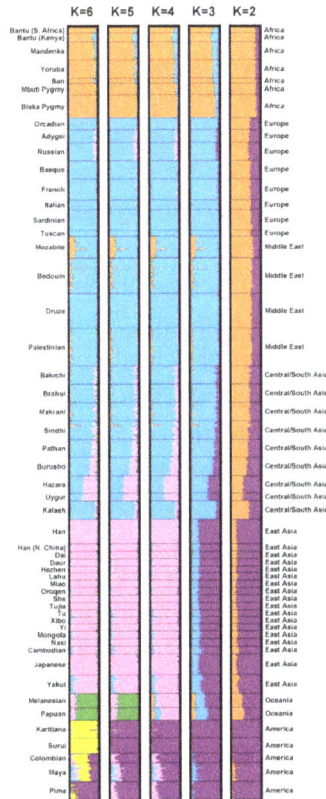

Human population structure can be inferred from multilocus DNA sequence data (Rosenberg et al. 2002, 2005). Individuals from 52 populations were examined at 993 DNA markers. This data was used to partition individuals into K = 2, 3, 4, 5, or 6 gene clusters. In this figure, the average fractional membership of individuals from each population is represented by horizontal bars partitioned into K colored segments.

Criticism

The Rosenberg study has been criticised on several grounds.

The existence of allelic clines and the observation that the bulk of human variation is continuously distributed, has led some scientists to conclude that any categorization schema attempting to partition that variation meaningfully will necessarily create artificial truncations. (Kittles & Weiss 2003). It is for this reason, Reanne Frank argues, that attempts to allocate individuals into ancestry groupings based on genetic information have yielded varying results that are highly dependent on methodological design. Serre and Pääbo (2004) make a similar claim:

The absence of strong continental clustering in the human gene pool is of practical importance. It has recently been claimed that "the greatest genetic structure that exists in the human population occurs at the racial level" (Risch et al. 2002). Our results show that this is not the case, and we see no reason to assume that "races" represent any units of relevance for understanding human genetic history.

In a response to Serre and Pääbo (2004), Rosenberg *et al.* (2005) maintain that their clustering analysis is robust. Additionally, they agree with Serre and Pääbo that membership of multiple clusters can be interpreted as evidence for clinality (isolation by distance), though they also comment that this may also be due to admixture between neighbouring groups (small island model). Thirdly they comment that evidence of clusterdness is not evidence for any concepts of "biological race".

Clustering does not particularly correspond to continental divisions. Depending on the parameters given to their analytical program, Rosenberg and Pritchard were able to construct between divisions of between 4 and 20 clusters of the genomes studied, although they excluded analysis with more than 6 clusters from their published article. Probability values for various cluster configurations varied widely, with the single most likely configuration coming with 16 clusters although other 16-cluster configurations had low probabilities. Overall, "there is no clear evidence that K=6 was the best estimate" according to geneticist Deborah Bolnick (2008:76-77). The number of genetic clusters used in the study was arbitrarily chosen. Although the original research used different number of clusters, the published study emphasized six genetic clusters. The number of genetic clusters is determined by the user of the computer software conducting the study. Rosenberg later revealed that his team used pre-conceived numbers of genetic clusters from six to twenty "but did not publish those results because Structure [the computer program used] identified multiple ways to divide the sampled individuals". Dorothy Roberts, a law professor, asserts that "there is nothing in the team's findings that suggests that six clusters represent human population structure better than ten, or fifteen, or twenty." When instructed to find two clusters, the program identified two populations anchored around by Africa and by the Americas. In the case of six clusters, the entirety of Kalesh people, an ethnic group living in Northern Pakistan, was added to the previous five.

Commenting on Rosenberg's study, law professor Dorothy Roberts wrote that "the study actually showed that there are many ways to slice the expansive range of human genetic variation.

Clusters in Tishkoff et al. 2009

Sarah A. Tishkoff and colleagues analyzed a global sample consisting of 952 individuals from the HGDP-CEPH survey, 2432 Africans from 113 ethnic groups, 98 African Americans, 21 Yemenites, 432 individuals of Indian descent, and 10 Native Australians. A global STRUCTURE analysis of these individuals examined 1327 polymorphic markers, including of 848 STRs, 476 indels, and 3 SNPs. The authors reported cluster results for K=2 to K=14. Within Africa, six ancestral clusters were inferred through Bayesian analysis, which were closely linked with ethnolinguistic heritage. Bantu populations grouped with other Niger-Congo-speaking populations from West Africa. African Americans largely belonged to this Niger-Congo cluster, but also had significant European ancestry. Nilo-Saharan populations formed their own cluster. Chadic populations clustered with the Nilo-Saharan groups, suggesting that most present-day Chadic speakers originally spoke languages from the Nilo-Saharan family and later adopted Afro-Asiatic languages. Nilotic populations from the African Great Lakes largely belonged to this Nilo-Saharan cluster too, but also had some Afro-Asiatic influence due to assimilation of Cushitic groups over the last 3,000 years. Khoisan populations formed their own cluster, which grouped closest with the Pygmy cluster. The Cape Coloured showed assignments from the Khoisan, European and other clusters due to the population's mixed heritage. The Hadza and Sandawe populations formed their own cluster. An Afro-Asiatic cluster was also discerned, with the Afro-Asiatic speakers from North Africa and the

Horn of Africa forming a contiguous group. Afro-Asiatic speakers in the Great Lakes region largely belonged to this Afro-Asiatic cluster as well, but also had some Bantu and Nilotic influence due to assimilation of adjacent groups over the last 3,000 years. The remaining inferred ancestral clusters were associated with European, Middle Eastern, Oceanian, Indian, Native American and East Asian populations.

Examining Effects of Sampling in Xing Et Al. 2010

Jinchuan Xing and colleagues used an alternate dataset of human genotypes including HapMap samples and their owns samples from a total of XX populations distributed roughly evenly across the Earth's land surface. They found that the alternate sampling reduced the F_{ST} estimate of inter-population differences from 0.18 to 0.11, suggesting that the higher number may be an artifact of uneven sampling. They conducted a cluster analysis using the ADMIXTURE program and found that "genetic diversity is distributed in a more clinal pattern when more geographically intermediate populations are sampled."

HUGO Asian Study

A study by the HUGO Pan-Asian SNP Consortium in 2009 using the similar principal components analysis found that East Asian and South-East Asian populations clustered together, and suggested a common origin for these populations. At the same time they observed a broad discontinuity between this cluster and South Asia, commenting "most of the Indian populations showed evidence of shared ancestry with European populations". It was noted that "genetic ancestry is strongly correlated with linguistic affiliations as well as geography".

Controversy of Genetic Clustering and Associations with "Race"

Studies of clustering reopened a debate on the scientific reality of race, or lack thereof. In the late 1990s Harvard evolutionary geneticist Richard Lewontin stated that "no justification can be offered for continuing the biological concept of race. (...) Genetic data shows that no matter how racial groups are defined, two people from the same racial group are about as different from each other as two people from any two different racial groups. This view has been affirmed by numerous authors and the American Association of Physical Anthropologists since. A.W.F. Edwards as well as Rick Kittles and Jeffrey Long have criticized Lewontin's methodology. Edwards also charged charged that Lewontin made an "unjustified assault on human classification, which he deplored for social reasons." In their 2015 article, Keith Hunley, Graciela Cabana, and Jeffrey Long recalculate the apportionment of human diversity using a more complex model than Lewontin and his successors. They conclude: "In sum, we concur with Lewontin's conclusion that Western-based racial classifications have no taxonomic significance, and we hope that this research, which takes into account our current understanding of the structure of human diversity, places his seminal finding on firmer evolutionary footing."

Genetic clustering studies, and particularly the five-cluster result published by Rosenberg's team in 2002, have been interpreted by journalist Nicholas Wade, evolutionary biologist Armand Marie Leroi, and others as demonstrating the biological reality of race. For Leroi, "Race is merely a shorthand that enables us to speak sensibly, though with no great precision, about genetic rather than cultural or political differences." He states that, "One could sort the world's population into 10, 100, perhaps

1,000 groups," and describes Europeans, Basques, Andaman Islanders, Ibos, and Castillians each as a "race." In response to Leroi's claims, the Social Science Research Council convened a panel of experts to discuss race and genomics online. In their 2002 and 2005 papers, Rosenberg and colleagues disagree that their data implies the biological reality of race. Over one hundred senior population geneticists denounced Wade's book *A Troublesome Inheritance* for misinterpreting their work.

In 2006, Lewontin wrote that any genetic study requires some priori concept of race or ethnicity in order to package human genetic diversity into defined, limited number of biological groupings. Informed by geneticist, zoologists have long discarded the concept of race for dividing up groups of non-human animal populations within a species. Defined on varying criteria, in the same species widely varying number of races could be distinguished. Lewontin notes that genetic testing revealed that "because so many of these races turned out to be based on only one or two genes, two animals born in the same litter could belong to different 'races'".

Studies that seek to find genetic clusters are only as informative as the populations they sample. For example, Risch and Burchard relied on two or three local populations from five continents, which together were supposed to represent the entire human race. Another genetic clustering study used three sub-Saharan population groups to represent Africa; Chinese, Japanese, and Cambodian samples for East Asia; Northern European and Northern Italian samples to represent "Caucasians". Entire regions, subcontinents, and landmasses are left out of many studies. Furthermore, social geographical categories such "East Asia" and "Caucasians" were not defined. "A handful of ethnic groups to symbolize an entire continent mimic a basic tenet of racial thinking: that because races are composed of uniform individuals, anyone can represent the whole group" notes Roberts.

The model of Big Few fails when including overlooked geographical regions such as India. The 2003 study which examined fifty-eight genetic markers found that Indian populations had their ancestral lineages to Africa, Central Asia, Europe, and southern China. Reardon, from Princeton University, asserts that flawed sampling methods are built into many genetic research projects. The Human Genome Diversity Project (HGDP) relied on samples which were assumed to be geographically separate and isolated. The relatively small sample sizes of indigenous populations for the HGDP do not represent the human species' genetic diversity, nor do they portray migrations and mixing population groups which has been happening since prehistoric times. Geographic areas such as the Balkans, the Middle East, North and East Africa, and Spain are seldom included in genetic studies. East and North African indigenous populations, for example, are never selected to represent Africa because they do not fit the profile of "black" Africa. The sampled indigenous populations of the HGDP are assumed to be "pure"; the law professor Roberts claims that "their unusual purity is all the more reason they cannot stand in for all the other populations of the world that marked by intermixture from migration, commerce, and conquest."

King and Motulsky, in a 2002 Science article, states that "While the computer-generated findings from all of these studies offer greater insight into the genetic unity and diversity of the human species, as well as its ancient migratory history, none support dividing the species into discrete, genetically determined racial categories". Cavalli-Sforza asserts that classifying clusters as races would be a "futile exercise" because "every level of clustering would determine a different population and there is no biological reason to prefer a particular one." Bamshad, in 2004 paper published in Nature, asserts that a more accurate study of human genetic variation would use an objective sampling method. An objective sampling method would chose populations randomly and systematically across the world, including those populations which are characterized by historical intermingling, instead of cherry-picking population samples which fit a priori concept of racial classification. Roberts states that "if research collected DNA samples continuously from region to region throughout the world, they would find it impossible to infer neat boundaries between large geographical groups."

Anthropologists such as C. Loring Brace, philosophers Jonathan Kaplan and Rasmus Winther, and geneticist Joseph Graves, have argued that while there it is certainly possible to find biological and genetic variation that corresponds roughly to the groupings normally defined as "continental races", this is true for almost all geographically distinct populations. The cluster structure of the genetic data is therefore dependent on the initial hypotheses of the researcher and the populations sampled. When one samples continental groups the clusters become continental, if one had chosen other sampling patterns the clustering would be different. Weiss and Fullerton have noted that if one sampled only Icelanders, Mayans and Maoris, three distinct clusters would form and all other populations could be described as being clinally composed of admixtures of Maori, Icelandic and Mayan genetic materials. Kaplan and Winther therefore argue that seen in this way both Lewontin and Edwards are right in their arguments. They conclude that while racial groups are characterized by different allele frequencies, this does not mean that racial classification is a natural taxonomy of the human species, because multiple other genetic patterns can be found in human populations that crosscut racial distinctions. Moreover, the genomic data underdetermines whether one wishes to subdivisions (i.e., splitters) or a continuum (i.e., lumpers). Under Kaplan and Winther's view, racial groupings are objective social constructions that have conventional biological reality only insofar as the categories are chosen and constructed for pragmatic scientific reasons.

Genetic clustering was also criticized by Penn State anthropologists Kenneth Weiss and Brian Lambert. They asserted that understanding human population structure in terms of discrete genetic clusters misrepresents the path that produced diverse human populations that diverged from shared ancestors in Africa. Ironically, by ignoring the way population history actually works as one process from a common origin rather than as a string of creation events, structure analysis that seems to present variation in Darwinian evolutionary terms is fundamentally non-Darwinian."

Commercial Ancestry Testing and Individual Ancestry

Commercial ancestry testing companies, who use genetic clustering data, have been also heavily criticized. Limitations of genetic clustering are intensified when inferred population structure is applied to individual ancestry. The type of statistical analysis conducted by scientists translates poorly into individual ancestry because they are looking at difference in frequencies, not absolute differences between groups. Commercial genetic genealogy companies are guilty of what Pillar Ossorio calls the "tendency to transform statistical claims into categorical ones". Not just individuals of the same local ethnic group, but two siblings may end up beings as members of different continental groups or "races" depending on the alleles they inherit.

Many commercial companies use data from the International HapMap Project (HapMap)'s initial phrase, where population samples were collected from four ethnic groups in the world: Han Chinese, Japanese, Yoruba Nigerian, and Utah residents of Northern European ancestry. If a person has ancestry from a region where the computer program does not have samples, it will compensate with the closest sample that may have nothing to do with the customer's actual ancestry: "Consider a genetic ancestry testing performed on an individual we will call Joe, whose eight great-grandparents were from southern Europe. The HapMap populations are used as references for testing Joe's genetic ancestry. The HapMap's European samples consist of "northern" Europeans. In regions of Joe's genome that vary between northern and southern Europe (such regions might include the lactase gene), the genetic ancestry test is using the HapMap reference population is likely to incorrectly assign the ancestry of that portion of the genome to a non-European population because that genomic region will appear to be more similar to the HapMap's Yoruba or Han Chinese samples than to Northern European samples. Likewise, a person having Western European and Western African ancestries may have ancestors from Western Europe and West Africa, or instead be assigned to East Africa where various ancestries can be found. "Telling customers that they are a composite of several anthropological groupings reinforces three central myths about race: that there are pure races, that each race contains people who are fundamentally the same and fundamentally different from people in other races, and that races can be biologically demarcated." Many companies base their findings on inadequate and unscientific sampling methods. Researchers have never sampled the world's populations in a systematic and random fashion.

Geographical and Continental Groupings

Roberts argues against the use of broad geographical or continental groupings: "molecular geneticists routinely refer to African ancestry as if everyone on the continent is more similar to each other than they are to people of other continents, who may be closer both geographically and genetically. Ethiopians have closer genetic affinity with Armenians and Norwegians than with Bantu populations. Similarly, Somalis are genetically more similar to Gulf Arab populations than to other populations in Africa. Braun and Hammonds (2008) asserts that the misperception of continents as natural population groupings is rooted in the assumption that populations are natural, isolated, and static. Populations came to be seen as "bounded units amenable to scientific sampling, analysis, and classification". Human beings are not naturally organized into definable, genetically cohesive populations.

References

- Wade, Nicholas (2015-04-28). A Troublesome Inheritance: Genes, Race and Human History. Penguin. ISBN 978-0-14-312716-1.

- Raff, Jennifer (2014-07-01). "Nicholas Wade and Race: Building a Scientific Façade". Human Biology. 86 (3): 227–232. doi:10.13110/humanbiology.86.3.0227. ISSN 0018-7143. Retrieved 2016-06-24.

- Edwards, A.W.F. (2003-08-01). "Human genetic diversity: Lewontin's fallacy". BioEssays. 25 (8): 798–801. doi:10.1002/bies.10315. ISSN 1521-1878. PMID 12879450. Retrieved 2016-06-11.

- Long, Jeffrey C.; Kittles, Rick A. (2009). "Human Genetic Diversity and the Nonexistence of Biological Races". Human Biology. 81 (5): 793–794. doi:10.3378/027.081.0621. ISSN 1534-6617. Retrieved 2016-01-13.

- Hunley, Keith L.; Cabana, Graciela S.; Long, Jeffrey C. (2015-12-01). "The apportionment of human diversity revisited". American Journal of Physical Anthropology. 160: –. doi:10.1002/ajpa.22899. ISSN 1096-8644. Retrieved 2016-01-23.

- Lewontin, R. C. (1972). Theodosius Dobzhansky, Max K. Hecht, William C. Steere (eds.). "The Apportionment of Human Diversity". Evolutionary Biology. 6: 381–398. doi:10.1007/978-1-4684-9063-3_14. Retrieved 2016-06-09.

• Barbujani, G.; Ghirotto, S.; Tassi, F. (2013-09-01). "Nine things to remember about human genome diversity". Tissue Antigens. 82 (3): 155–164. doi:10.1111/tan.12165. ISSN 1399-0039. Retrieved 2015-12-02.

• Quote from Rosenberg, Noah A.; Pritchard, Jonathan K.; Weber, James L.; Cann, Howard M.; Kidd, Kenneth K.; Zhivotovsky, Lev A.; Feldman, Marcus W. (2002-12-20). "Genetic Structure of Human Populations". Science. 298 (5602): 2381–2385. doi:10.1126/science.1078311. ISSN 0036-8075. PMID 12493913. Retrieved 2015-09-15.

Permissions

Index

www.ingramcontent.com/pod-product-compliance
Lightning Source LLC
Chambersburg PA
CBHW082048190326
41458CB00010B/3485